International Max Planck Research School (IMPRS)
for Maritime Affairs
at the University of Hamburg

For further volumes:
http://www.springer.com/series/6888

Hamburg Studies on Maritime Affairs
Volume 25

Anja Rösel

Detection of Melt Ponds on Arctic Sea Ice with Optical Satellite Data

Springer

Anja Rösel
Institute of Oceanography
University of Hamburg
Hamburg
Germany

Erstgutachter: Prof. Dr. Lars Kaleschke
Zweitgutachter: Prof. Dr. Detlef Stammer
Tag der mündlichen Prüfung: 28.06.2012

ISSN 1614-2462 ISSN 1867-9587 (electronic)
ISBN 978-3-642-37032-8 ISBN 978-3-642-37033-5 (eBook)
DOI 10.1007/978-3-642-37033-5
Springer Heidelberg New York Dordrecht London

Library of Congress Control Number: 2013939063

Printed on acid-free paper

Springer is part of Springer Science+Business Media (www.springer.com)

Dedicated to Mats and Jens

Preface

This book is based on my dissertation which I wrote at the Institute of Oceanography of the University of Hamburg under the supervision of Prof. Dr. Lars Kaleschke. The dissertation was accepted by the Department of Geosciences in 2012.

I started my work in November 2008 with the intention to use the now freely accessible huge Landsat satellite imagery archive to detect melt ponds on the Arctic sea ice surface. It turned out that this archive consists mainly of scenes covering landscape and not the ocean surface. Additionally, most of the few available scenes were not usable for a multispectral classification due to over-saturation of the sensor. Therefore I had to look for alternative sensors. I decided to use aquisitions of the MODIS sensor—they have a coarser spatial resolution than those from the Landsat sensor, but provide a full coverage of the Arctic region on a daily basis.

Despite the technical problems at the beginning of my studies, I was able to submit my dissertation in March 2012. My work was enabled by a scholarship of the International Max Planck Research School for Maritime Affairs.

Hamburg, Germany
December 2012

Anja Rösel

Acknowledgments

I thank my academic advisor Prof. Dr. Lars Kaleschke for his supervision and constructive discussions during my work at the Institute of Oceanography at the University of Hamburg. I am grateful for the opportunity to develop my Ph.D. thesis in his sea ice group.

My special thanks are given to all the actual and former members contributing to this group for their willingness to discuss my work and to provide scientific, technical, and personal support.

My thanks go namely to my colleagues and friends Antje, Nina, Steffen, Freddy, Iris, Meike D., Meike S., Stefan, Marketa, Martin S., Martin G., Alex, David, Remon, Xiangshan, Amelie, and Frauke for their collegiality, friendship, and the cheerful time we spent at the institute and in our spare time.

I thank Donald K. Perovich from CRREL for providing the HOTRAX melt pond data and Florence Fetterer from NSIDC for her efforts in providing more melt pond statistics. Gerit Birnbaum, Wolfgang Dierking, Christof Lüpkes, and Jörg Hartmann from the AWI participated in the MELTEX aircraft campaign in 2008 and provided useful information for our study.

This work was funded by a scholarship of the International Max Planck Research School for Maritime Affairs in Hamburg. Additional support was provided by the Cluster of Excellence CliSAP (EXC177), University of Hamburg, funded through the German Science Foundation (DFG).

Finally, I want to give special thanks to my family and to my husband for their love and encouragement.

Contents

Acronyms

ANN	Artificial neural network
AWI	Alfred Wegener Institute for Polar and Marine Research, Bremerhaven, Germany—http://www.awi.de/
BFGS	Broyden–Fletcher–Goldfarb–Shanno algorithm
BRDF	Bidirectional reflectance distribution function
CALIPSO	Cloud–Aerosol Lidar and Infrared Pathfinder Satellite Observations
CliSAP	Integrated Climate System Analysis and Prediction
CRREL	Cold Regions Research and Engineering Laboratory
ETM+	Enhanced Thematic Mapper Plus
gdal	Geospatial Data Abstraction Library—http://www.gdal.org/
HDF	Hierarchical Data Format—http://www.hdfgroup.org/
HOTRAX	Healy-Oden Trans-Arctic Expedition 2005
MELTEX	Impact of melt ponds on energy and momentum fluxes between atmosphere and sea ice
MODIS	Moderate Resolution Imaging Spectroradiometer
NetCDF	Network Common Data Form
NSIDC	The National Snow and Ice Data Center, Boulder, Colorado—http://nsidc.org/
PCA	Principal component analysis
RMSE	Root Mean Square Error
SLC	Scan line corrector
SPOT	Satellite Pour l'Observation de la Terre
TOA	Top-of-the-atmosphere
UTM	Universal Transverse Mercator coordinate system

Abstract

The Arctic sea ice is characterized by profound changes caused by surface melting processes and the formation of melt ponds in summer. Melt ponds contribute to the ice-albedo feedback as they reduce the surface albedo of sea ice and hence accelerate the decay of Arctic sea ice. To quantify the melting on the entire Arctic sea ice, satellite-based observations are necessary.

Due to different spectral properties of snow, ice, and water multi-spectral optical sensors like Landsat 7 Enhanced Thematic Mapper plus (ETM+) or Moderate Resolution Image Spectroradiometer (MODIS) are theoretically applicable for the analysis of these distinct surface types. In this study the potential of both mentioned sensors to detect melt ponds on Arctic sea ice is demonstrated.

To analyze Landsat 7 ETM+ scenes, we have developed an algorithm based on principal component analysis (PCA) of two spectral channels for achieving the melt pond fraction. PCA allows differentiation of melt ponds and other surface types like snow, ice, or water. We use spectral bands 1 and 4 with central wavelengths at 480 and 770 nm, respectively. They represent best the differences in the spectral albedo of melt ponds. Due to problems caused by saturation of the sensor, it is only possible to analyze a few selected Landsat scenes which are not affected from oversaturation. A case study on a Landsat 7 ETM+ scene from July 19, 2001 using PCA was successfully performed.

For the analysis of MODIS data, we derived the melt pond fraction by using a spectral unmixing algorithm. This algorithm consists of a system of linear equations and was solved by an approximation method. Additionally, an artificial neural network was trained to reduce computational costs. Arctic-wide melt pond fractions and sea ice concentrations have been derived from the level 3 MODIS surface reflectance product. The validation of the MODIS melt pond data set has been conducted (i) with aerial photos from the MELTEX campaign 2008 in the Beaufort Sea, (ii) with very high resolution satellite data sets from the National Snow and Ice Data Center (NSIDC) for the years 2000 and 2001 representing four sites spread over the entire Arctic, and (iii) with ship observations from the trans-Arctic HOTRAX cruise in 2005. The root mean square errors (RMSE) range from 3.8 % for the comparison with HOTRAX data and 10.7 % for the comparison with NSIDC data to 10.3 % and 11.4 % for the comparison with MELTEX data with coefficients of determination ranging from $R^2 = 0.28$ to $R^2 = 0.45$, respectively.

For the first time, we have used satellite observations to analyze how melt pond fractions in the Arctic have developed spatially and temporally in the years 2000–2011. During this period the total melt pond area exhibits a significant negative trend of −16.4 %, corresponding to a declining sea ice extent, whereas the annual average of the relative melt pond fraction remains constant with 25.1 ± 1.7 % through the twelve melt cycles. Looking more into detail, a significant anomaly of the relative melt pond fraction at the beginning of the melt season in June 2007 followed by above-average values throughout the entire summer are documented. In contrast, the increase of the melt pond fraction at the beginning of June 2011 is within average values, but from mid-June 2011, relative melt pond fraction exhibits values up to two standard deviations above the mean values of 30.0 ± 1.2 % which are even higher than in summer 2007.

Furthermore, the influence of surface melting on sea ice concentration retrievals from passive microwave sensors is pointed out. The sea ice concentrations calculated from passive microwave data are up to 40 % lower compared to the results of the MODIS sea ice concentration.

Chapter 1
Introduction

Over the last three decades, significant reductions in sea ice extent and thickness of Arctic sea ice have been documented (e.g. [37, 46, 75]). A series of extreme September sea ice extent minima have been observed in the last 10 years, which have accelerated the rate of decline [76]. It is to assume that a growing importance of the ice-albedo feedback[1] has in part been attributed to the continued decline [13,58,76]. The ice-albedo feedback is a positive feedback process: decreasing surface albedo due to a melting ice cover leads to more absorption of solar radiation, resulting in additional melt processes. These melt processes again reduce the surface albedo and thus enhance melting of the ice cover [13].

In boreal summer, melt ponds are a common feature on Arctic sea ice and cover up to 50–60 % of the sea ice area [16, 21]. Melt ponds or surface melt puddles (Fig. 1.1) are defined as an accumulation of melt water on sea ice, mainly due to melting snow, but in more advanced stages also due to melting of sea ice. The distribution on the ice, the size and the depth of the ponds, as well as the color are very variable and depend on topography, surface, and atmospheric conditions. On a flat topography of first-year ice the melt pond fraction can rise up to 90 % [57].

The existence of melt ponds on Arctic sea ice causes a decrease of the surface albedo from approximately 0.8 to 0.5 due to a higher absorption of the incoming radiation. This effect initiates additional heat uptake [13,60]. Therefore, melt ponds have a significant influence on the rate of sea ice melt [61, 81] and on Earth's radiation balance [46, 50, 58], and the potential loss of a multi-year ice coverage [11,37,46,58,69].

In order to better constrain the role of sea ice for the Arctic amplification and Earth's climate system, it is important to quantify the overall distribution of melt ponds (e.g. [18, 30, 35, 52, 59, 69, 70, 77]). This is only possible with observations on a large scale over at least one melting period. Therefore, it is important to

[1] Technical terms are explained in the glossary. This is also valid for following expressions.

A. Rösel, *Detection of Melt Ponds on Arctic Sea Ice with Optical Satellite Data*, Hamburg Studies on Maritime Affairs 25, DOI 10.1007/978-3-642-37033-5_1,

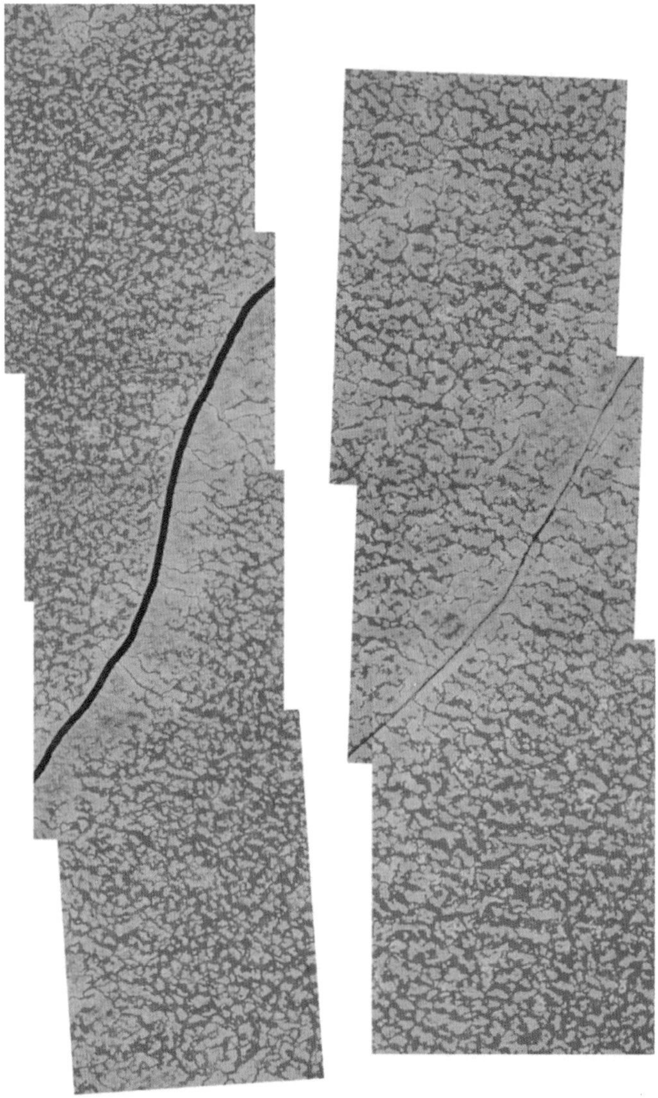

Fig. 1.1 Two short tracks of aerial photos, displaying Arctic sea ice covered with melt ponds. Some of them are draining into the nearby cracks. Photos were made during the MELTEX aircraft campaign on June 7, 2008 over Franklin Bay. Photos: Gerit Birnbaum, AWI

utilize satellite data and remote sensing techniques that are applicable to detect the evolution of melt ponds.

Literature values of spectral and total albedo for various Arctic surface types, acquired on field campaigns, range from 0.06 for open water, and 0.29 for mature melt ponds, to 0.87 for new snow (e.g. [6, 26, 27, 54, 56, 83]). Several field experiments and ship observations have been conducted on different locations in the Arctic Ocean to study albedo and spectral behavior of melt ponds [49, 56], as well as distribution and size of the ponds [19, 50, 55, 61, 67].

Although the potential of different optical properties of ponded sea ice in comparison to bare or snow-covered sea ice for satellite applications was proposed already by Grenfell in 1977 [26], the first implementations were published 20 years later. These studies are based on aerial video data [78], optical satellite data [44,63,64,79,81], and radar data [31,87]. Until now, all studies show only by way of example, that melt pond identification from different satellite data is possible. While these studies discuss the feasibility of deriving melt pond fractions from satellite data for specific examples, to our knowledge, no satellite-derived Arctic-wide, multi-annual melt pond data set exists.

Recently, a couple of new publications concerning the influence of melt ponds on the radiation budget were published [15,23,32,59], but all of them are either built on modeled melt pond data or result from local observations. This indicates the strong requirement of a large scale melt pond data set, that is based on observations.

In this study, the different spectral properties of sea ice surface types will be used to develop a technique to determine the melt pond fraction on the total Arctic sea ice area throughout the seasonal cycle from optical multi-spectral satellite sensors.

1.1 Objectives

The study area is the Arctic Ocean northward 60°N (see Fig. 1.2). The primary goal of this study is to provide a melt pond data set on a large spatial scale for a multi-annual time period. Additionally, this leads to the following research questions:

- What is the mean annual melt pond fraction of the Arctic and how does a seasonal cycle look like?
- Is there a relation of melt pond fraction and sea ice extent minimum?
- Do melt ponds influence sea ice concentration values as retrieved from passive microwave sensors?

To achieve the main goal and answer the research questions, we pursue the following strategy to quantify the melt pond fraction on Arctic sea ice:

Firstly, Landsat satellite data is used for this analysis, because of the very high spatial resolution of 30 m × 30 m. The full Landsat archive, covering a time period from 1972 until now, has been freely accessible since January 2009. A technique based on tie-points and spectral differences of two optical bands of a Landsat scene, which was published by Markus [44] is adapted to a different satellite scene. The results of this method are used as a basis for comparison with results from different techniques.

Secondly, to improve the proposed method of Markus [44], we develop a new technique, based on principal-component analysis (PCA), to identify the melt pond

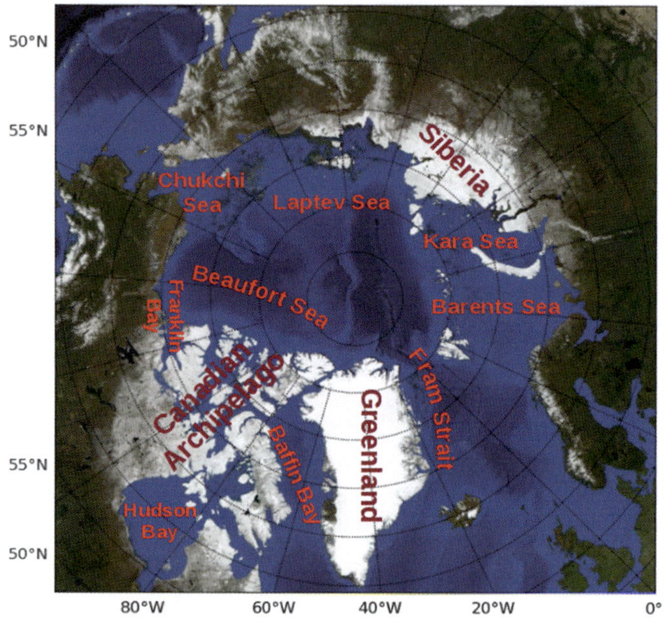

Fig. 1.2 Overview over the Arctic region

fraction from Landsat satellite data. In this study, it turns out, that the use of Landsat data is not applicable to create large data sets or time series, because the sensors have severe problems with observations over sea ice.

Thirdly, Moderate Resolution Image Spectroradiometer (MODIS) data with a resolution of 500 m × 500 m are used. Due to the coarser resolution compared to Landsat data, both above described methods cannot be applied to MODIS data. Hence, we modify a spectral unmixing algorithm, proposed by Tschudi [81], to obtain melt pond fractions from MODIS data by adding a side condition to limit the resulting surface fractions to values between zero and one. This modification causes high computational costs for the solution. To speed up processing, a trained artificial neural network (ANN) is used. With this procedure a multi-annual data set of melt pond fractions and sea ice concentrations for the entire Arctic is created for the first time.

For validation three different data sets are analyzed: firstly, results of the aircraft campaign MELTEX, conducted by the Alfred Wegener Institute for Polar and Marine Research (AWI) in May and June 2008 over the Beaufort Sea [5]. Secondly, sea ice melt pond statistics and maps of four Arctic Ocean sites during the summers of 2000 and 2001 from the National Snow and Ice Data Center (NSIDC) [22], and thirdly, results of the melt pond observations from the *Healy-Oden* Trans-Arctic Expedition 2005 (HOTRAX 2005) [55].

1.2 Structure

My thesis is structured as follows: after descriptions of the physical properties of sea ice, spectral differences of melt ponds, and melting processes on sea ice in Chap. 2, a brief introduction of the used sensors and the corresponding data sets are given (Chap. 3). Chapter 4 presents two different methods for melt pond determination from Landsat satellite data. This is followed by the description of a method for melt pond determination using MODIS data in Chap. 5. Finally, a summary of the thesis and an outlook are given in Chap. 6.

1.3 Publications

This thesis is based on the following publications, which have been published during my Ph.D. studies:

- Rösel, A. & Kaleschke, L. (2011). Comparison of different retrieval techniques for melt ponds on Arctic sea ice from Landsat and MODIS satellite data. *Annals of Glaciology*, **52(57)**. [63]
- Rösel, A., Kaleschke, L., & Birnbaum, G. (2012). Melt ponds on Arctic sea ice determined from MODIS satellite data using an neuronal network. *The Cryosphere*, **6-2**. [65]
- Rösel, A., & Kaleschke, L. (2012). Exceptional melt pond occurrence in the years 2007 and 2011 on the Arctic sea ice revealed from MODIS satellite data. *JGR*, **117**. [64]

Excerpts of these publications will not be cited anymore in the following thesis.

Chapter 2
Physical Characteristics of Sea Ice

Perovich declares in his monograph "The Optical Properties of Sea Ice" [54] parameters which are used to describe reflection, absorption, and transmission of sea ice (see Fig. 2.1). In the recent study, the main focus is on the definitions related to the reflection of solar incidence on sea ice.

2.1 Definitions

The spectral radiance $I(\theta, \phi, \lambda)$ is the power of solar light in a particular direction, defined by the zenith angle θ, and the azimuth angle ϕ at a particular wavelength λ. The spectral radiance has units of W m^{-2} sr^{-1} nm^{-1} [54].

The spectral irradiance $F(\lambda)$ is the radiance projected onto a plane surface and integrated over a hemisphere. Because of this projection the radiance is scaled by $\cos\theta$. The downwelling irradiance $F_d(\lambda)$ is the radiance integrated over downward directions like the sky, and the upwelling irradiance $F_u(\lambda)$ is the radiance integrated over upward directions [54]. This can be expressed as:

$$F_d(\lambda) = \int_{\phi=0}^{2\pi} \int_{\theta=0}^{\pi/2} I(\theta, \phi, \lambda) \cos\theta \sin\theta d\theta d\phi \tag{2.1}$$

$$F_u(\lambda) = \int_{\phi=0}^{2\pi} \int_{\theta=\pi/2}^{\pi} I(\theta, \phi, \lambda) \cos\theta \sin\theta d\theta d\phi. \tag{2.2}$$

Hence, the albedo (α) at a particular wavelength (λ) for the surface can be described as fraction of the incident irradiance that is reflected:

A. Rösel, *Detection of Melt Ponds on Arctic Sea Ice with Optical Satellite Data*, Hamburg Studies on Maritime Affairs 25, DOI 10.1007/978-3-642-37033-5_2,

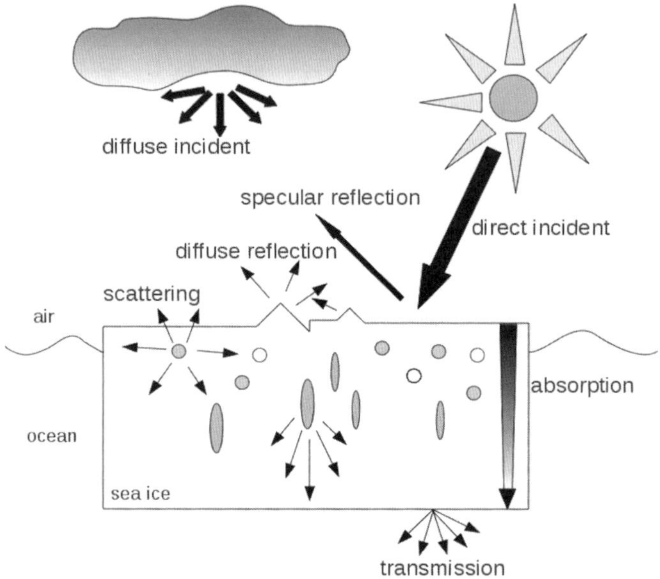

Fig. 2.1 Scheme of radiative transfer in sea ice (modified from Perovich, 1996)

$$\alpha(\lambda) = \frac{F_u(\lambda)}{F_d(\lambda)}, \tag{2.3}$$

where $\alpha(\lambda)$ is named as spectral albedo or spectral reflectance value.

For many applications the wavelength-integrated or total albedo (α_t) is used, since it is a measure of the total solar energy absorbed by the surface [26, 56]. In the following context, when only albedo is mentioned, we typically refer to total albedo. The total albedo can be expressed as:

$$\alpha_t = \frac{\int_{300}^{3000} \alpha(\lambda) F_d(\lambda) d\lambda}{\int_{300}^{3000} F_d(\lambda) d\lambda}, \tag{2.4}$$

where the integral interval covers the incident solar spectrum from 300 to 3,000 nm.

Due to varying incident radiation, total albedo is dependent on sky conditions [26, 28, 83] as well as the solar zenith angle [29, 83]. Under cloudy conditions with high solar zenith angles, the total albedo of the same surface type is higher than under clear sky conditions with low solar zenith angles [28, 29]. Perovich describes that overcast albedo values can be up to 0.1 higher than clear sky albedo values [54] (see Table 2.1). In this recent study, satellite data from optical sensors are used which can only be obtained under clear conditions. Therefore, all values given in the result section can be assumed as clear sky albedo values.

Table 2.1 Total albedo of different sea ice surfaces for clear-sky and cloudy conditions (after Grenfell [28])

Surface type	α[clear]	α[cloudy]	$\Delta\alpha$
Cold snow	0.851	0.929	0.077
Melting snow	0.607	0.678	0.071
Deteriorated melting ice	0.436	0.492	0.056
Undeterioated melting ice	0.294	0.332	0.037
Blue–green ice	0.273	0.309	0.036
Dirty ice	0.347	0.367	0.020
Early melt pond	0.239	0.266	0.027
Late melt pond	0.129	0.141	0.012

To calculate albedo on a large scale, surface-based total albedo values are weighted with the fraction of their corresponding surface component [21, 55, 81].

The so-called areal-averaged albedo $\overline{\alpha}$ can be expressed as

$$\overline{\alpha} = \alpha_l A_l + \alpha_p A_p + \alpha_i A_i , \tag{2.5}$$

where A is the area fraction, α is the wavelength-integrated albedo and l, p, i stand for leads, ponds, and ice, respectively. Total albedo values of different surface types can be found in the literature and are also listed in Sect. 2.3.

2.2 Surface Reflection

To describe surface reflection, it is essential to give definitions of the used angles: the zenith angles θ are defined as 90° minus elevation angle of the sun or sensor. The azimuth angles ϕ are the differences from the viewing direction from the sun or sensor to the North. The relative azimuth and zenith angles are the relative positions from sun or sensor to each other (see Fig. 2.2).

Surface reflection has variable characteristics (Fig. 2.3): The one extreme is the specular reflection, where the reflected angle θ_r is equal to the incident angle θ_i [62]. A calm water surface is a natural surface that represents most a specular reflector, where the upward reflected flux F_r is equal the incident flux F_i.

The other extreme is the Lambertian reflection: The flux of upward reflected radiation is equally distributed over all angles, irrespective of the direction of the source. This can be expressed as

$$F_r = \pi F_i . \tag{2.6}$$

Most surface types do not behave like a Lambertian or a specular reflector. Therefore, a continuously varying function of both, the incident direction of

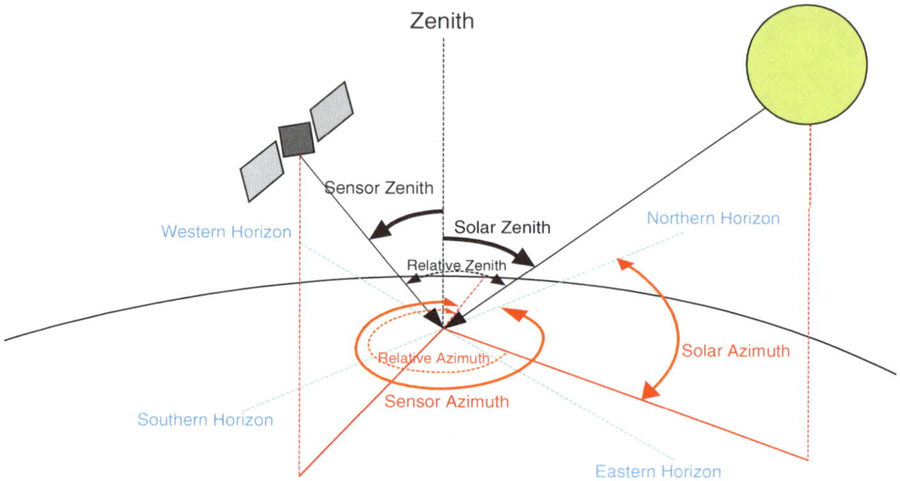

Fig. 2.2 Definition of solar zenith angle, sensors zenith angle, solar azimuth angle, sensor azimuth angle, relative azimuth angle, relative zenith angle (phase angle)

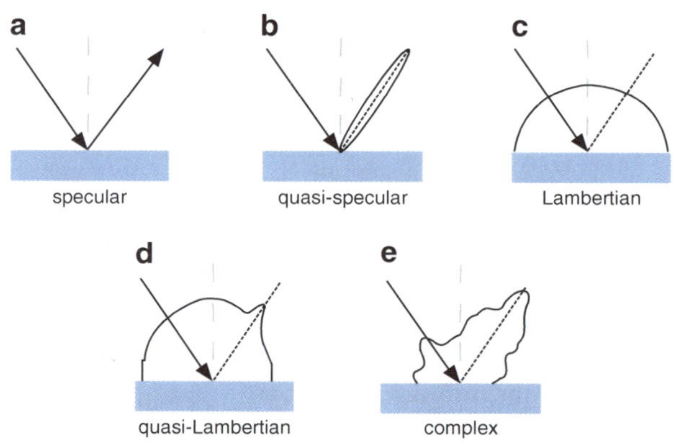

Fig. 2.3 Examples of various types of surface reflection (after Petty, 2006)

radiation, and the reflected direction is described as the bidirectional reflection distribution function (BRDF) $\rho(\theta_i, \phi_i, \theta_r, \phi_r)$ [83]. With the BRDF, the reflected intensity can be described as $I \uparrow$ of a surface in relation to a specific direction (θ_r, ϕ_r):

$$I \uparrow (\theta_r, \phi_r) = \rho(\theta_i, \phi_i, \theta_r, \phi_r) S_0 \cos\theta_i, \quad (2.7)$$

where S_0 is the solar flux normal to the beam. This relationship assumes direct illumination of the surface by the sun at position (θ_i, ϕ_i) in the sky. Therefore, this relationship is only valid for clear sky observations.

Fig. 2.4 Arctic sea ice surface covered with melt ponds displaying various characteristics. The photo was taken from a helicopter during the Polarstern cruise ARK-XXII/2 in 2007. Photo: Stefan Kern

Fig. 2.5 Arctic sea ice surface covered with light blue and dark blue melt ponds. An inhomogeneous surface composed of floes with melt pond and ridges and cracks, intersected by open water areas makes a surface classification with moderate resolution satellite sensors difficult. The photo was taken from a helicopter during the Polarstern cruise ARK-XXII/2 in 2007. Photo: Stefan Kern

2.3 Optical Properties of Sea Ice

Sea ice is a composite of small fractionated areas of melt ponds, leads, snow fields, and ridges on a scale of meters over tens of meters to hundreds of meters. This results in a very inhomogeneous surface (see Figs. 2.4–2.6).

Additionally, sea ice is composed of first-year and multi-year ice. Multi-year ice has survived at least one melt season. The structure as well as the brine and

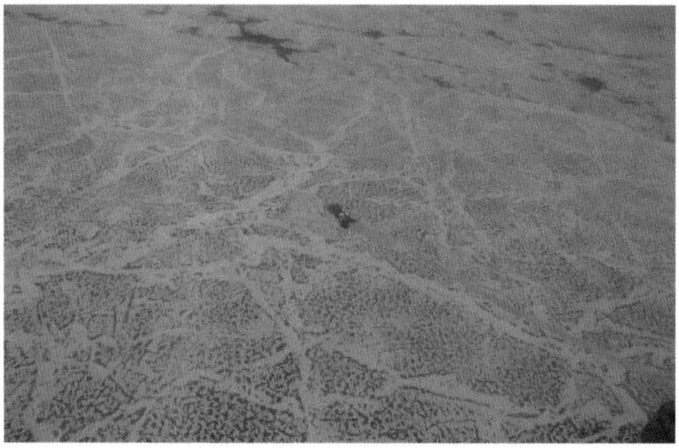

Fig. 2.6 Arctic sea ice surface covered with melt ponds. In the center of the photo is the research vessel Polarstern for comparison of the scales. The photo was taken from a helicopter during the Polarstern cruise ARK-XXII/2 in 2007. Photo: Stefan Kern

air inclusions are different for each type of sea ice [72]. Therefore, each of these categories has different physical and optical properties. By determining the albedo changes of a larger area during the melting season, the knowledge of the temporal and spatial albedo variability in each of these ice categories is essential [56].

The optical properties of ice and snow are a strong function of the wavelength of the incident solar radiation (Fig. 2.7 on page 13). Highest spectral albedo values (>0.9) appear in short wavelength ranges from 400 to 600 nm for dry snow. The spectral albedo decreases toward longer wavelengths at a rate which seems to be related to the liquid–water content of the surface layer [26]. At 500 nm melt ponds have albedo values that can range between 0.6 for young and shallow ponds and 0.25 for matured ponds on multi-year ice. The variety of albedo values for ponds is caused by differences in depths and underlying surfaces (see Figs. 2.8 and 2.9 on pages 14f.).

The albedo is correlated to the amount of air bubbles and brine inclusions in the sea ice. Hence, we can distinguish between albedo values of first-year ice and multi-year ice [54]: the albedo of first-year ice is generally higher than the albedo of multi-year ice.

It is recognized from Fig. 2.7, that snow covered ice, bare ice, and melting bare ice (a–c) show a smaller reduction in the spectral albedo at higher wavelengths than the both types of melt ponds (d and e). The spectral albedo of ponded ice is characterized by a decrease between 500 and 800 nm. Because of the relatively high transparency of water at short wavelengths, values below 500 nm are determined primarily by scattering properties of underlying ice. The transition zone (500–800 nm) represents a region where absorption of water becomes a dominant factor. Above 800 nm radiation is almost totally absorbed by the water surface and the underlying ice layer has no influence on the albedo [26]. These spectral

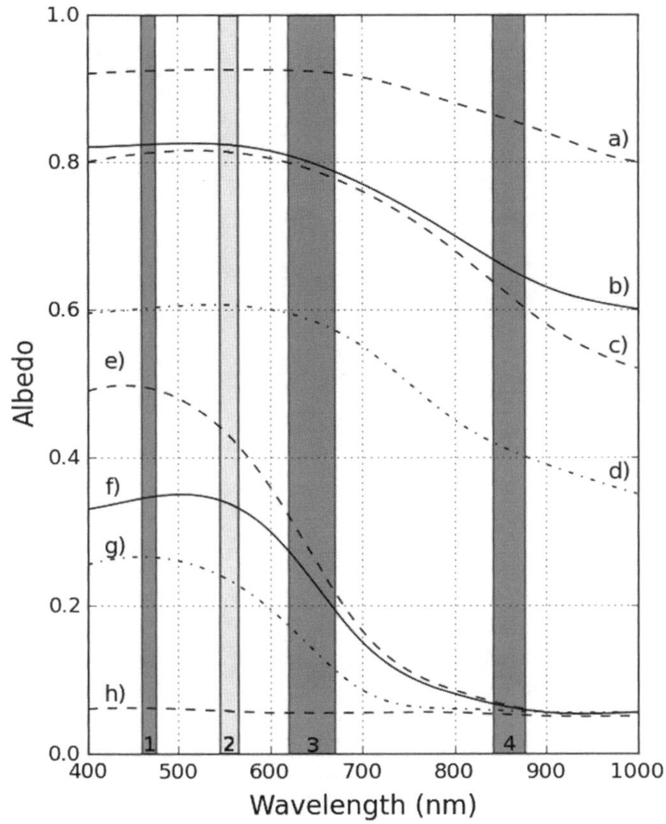

Fig. 2.7 Spectral albedo values for different surface types on Arctic sea ice: (*a*) snow-covered ice (dry snow), (*b*) cold bare ice, (*c*) wet snow, (*d*) melting first year ice, (*e*) young melt pond, (*f*) and (*g*) two types of mature melt ponds, and (*h*) open water. The gray columns represent the range of the first four MODIS bands. For this study the spectral bands 1, 3, and 4 are used [26]

differences of snow, ice, and melt ponds can be used to separate melt ponds from bare or snow covered ice [63].

In literature many spectral and total albedo values for different surface types are given [6,26,27,54,56,83]. A broad collection of albedo values from various sources is published on the website of the University of Alaska.[1] Figure 2.10 on page 14 shows some total albedo values published by Perovich [54]. The values are ranging from 0.06 for open water over 0.29 for mature melt ponds to 0.87 for new snow.

---

[1] http://www.gi.alaska.edu/~eicken/he_teach/GEOS615icenom/albedo/albedo%20classification.htm, accessed in February 2012.

Fig. 2.8 Arctic sea ice surface covered with mainly light blue melt ponds. This photo shows a temporal ice station on a floe for measuring ice thickness. The photo was taken from a helicopter during the Polarstern cruise ARK-XXII/2 in 2007. Photo: Stefan Kern

Fig. 2.9 Arctic sea ice surface covered with mainly dark melt ponds. This photo shows a temporal ice station on a flow for measuring ice thickness. The photo was taken from a helicopter during the Polarstern cruise ARK-XXII/2 in 2007. Photo: Stefan Kern

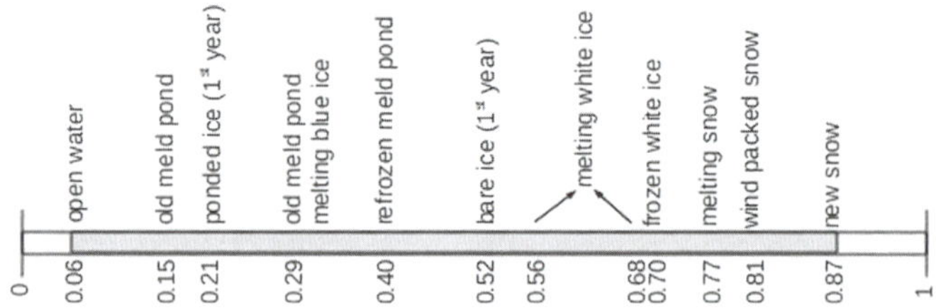

Fig. 2.10 Wavelength-integrated albedos for different surface types on Arctic sea ice [54]

2.4 Melting Processes on Sea Ice

To understand the development of sea ice throughout the melting period, it is necessary to distinguish between five distinct phases in the albedo evolution: dry snow, melting snow, pond formation, pond evolution, and refreezing [61].

In winter, most of the ice surface is covered with a dry snow layer of variable depth, building a more or less homogeneous surface with a high total albedo between 0.8 and 0.9 [61]. With onset of the summer melt season, the sea ice cover is subject to profound changes in its physical state and optical properties. The point in time when the melting process begins, strongly depends on the amount of solar energy absorbed before and during the melt season. It should be noted that early melt onset allows an earlier development of open water areas—which then again enhance the ice-albedo feedback [58]. Additionally, a zonal onset of the first melting processes is observable, starting in sub-Arctic regions like Bering Sea and Hudson Bay in mid-April. In the Central Arctic, first melting starts in June [45]. These observations are clearly related to the increasing solar radiation during spring and summer and have a strong latitudinal dependence. A trend to an earlier melt onset and a later freeze-up date in the entire Arctic region for the last three decades is described in literature [45]. The resulting longer melting periods are again a positive factor to the ice-albedo feedback mechanism.

Starting in April in sub-Arctic regions, dry snow wettens and begins to melt. Snow grains transform and grain size generally increases [28]. Even these first melting processes can reduce the albedo of snowy surfaces by about 10–20 % .

Snow melting processes depend on the properties of the snow cover, mainly on the snow depth. The variability of Arctic snow cover depth ranges from none to several meters in leeward sides of ridges or other obstacles. In the Central Arctic, snow cover usually disappears by end of June [21]. Melt water of snow and ice accumulates in surface depressions and other surface deformation features. Compared to the much more irregular surface topography of multi-year ice, plane and flat surfaces of first-year ice have the potential to host large and extended melt pond areas [16, 21]. They can reach a coverage of over 50 % of the total sea ice area [28]. On a flat topography of first-year level ice and in an early melt stage, the melt pond fraction can even rise up to 90 % [57]. As melting develops, pond water drains through porous ice and cracks [17]. The pond properties and distribution on multi-year ice are described as smaller, deeper, and more numerous than on first-year ice [88].

The heat transfer due to convection in water exceeds the one of ice. Additionally, the lower albedo of ponded ice allows a higher penetration of heat into the ice. Both factors yield to a two to three times higher melt rate beneath ponds compared to the melt rate of bare ice [21]. Hence, the ponds deepen and can even melt through the ice layer. With the increasing depth of the ponds, also the diameter decreases [21].

On the one hand, spectral as well as total albedo of bare ice are fairly constant during the melting period. On the other hand, albedo of ponded ice depends on the pond depth and varies throughout the melting period [56].

Fig. 2.11 Refrozen and partly snow covered melt ponds on Arctic sea ice. The photo was taken during the Polarstern cruise ARK-XXII/2 in 2007. Photo: Stefan Kern

Melt ponds are nearly salt free and the density maximum of the ponded water lies well above the freezing point [21]. Consequently, radiative heating favors convection within the pond: due to the density anomalies of water, the warmer water will sink down and thus causes further melting. Convection and mixing of the water is additionally enhanced by wind [17]. In late summer, melt ponds tend to melt down to sea level and drain towards the ocean.

Mature ponds are effective traps for the first drifting snow. Through the capillar effect, the water level of the pond rises. Therefore, it is less likely that this particular area will be pond covered in the next melting season [21]. Freeze-up starts in late August or early September, caused by low air temperatures. This results in a decreasing melt pond fraction. A snowfall event after freeze-up will cover the melt ponds and albedo resulting in a higher surface albedo (see Fig. 2.11).

The large inter-annual variability of the melt pond coverage can be caused by several factors: year-to-year variations of weather (mainly clouds and radiation), amount of melt water availability from snow, variable surface topography, floe size distribution, and their effect on runoff-pattern [21, 71].

Chapter 3
Optical Remote Sensing

The sensors used in this study are part of the on-board measurement system of the satellites Landsat 7, Terra and Aqua. All three are near-polar orbiting satellites. This ensures acquisition of a neighboring swath during the next orbit. Their orbits are sun-synchronous, resulting in image acquisition at nearly the same local time for each area of the world. For more technical details of satellites and sensors see [34].

3.1 Landsat 7 ETM+

The sensor Enhanced Thematic Mapper plus (ETM+) on board the satellite Landsat 7 is an opto-mechanical multi-spectral scanner, which scans the surface line by line with an oscillating mirror perpendicular to the flight direction (across-track scanning). Each line represents a 185 km long and 480 m wide sweep on the surface [38]. The inclination of the orbit is 98.2°. This allows acquisitions of data up to a latitude of 82°N.

The incoming reflected radiation passes a telescope and is separated into seven spectral bands with 30 m (reflective bands 1–5 and 7) and 60 m (thermal band 6) spatial resolution[1] with a range from 0.45 to 12.5 μm. The received electromagnetic energy is converted into 256 discrete levels referred to as digital numbers (DN). The ETM+ has an additional panchromatic band with 15 m spatial resolution. In contrast to SPOT or Terra, Landsat is not able to look at any position other than directly downward, therefore the revisiting period is fixed at 16 days. Getting closer to the pole, the overlap of adjacent orbit paths increases and the same point is acquired more often.

The scan line corrector (SLC), which compensates for the forward motion of the satellite, failed on May 31, 2003 [38]. The resulting effect is a "striped" image with

---

[1]Detailed wavelength ranges are listed in Appendix A.1.

A. Rösel, *Detection of Melt Ponds on Arctic Sea Ice with Optical Satellite Data*, Hamburg Studies on Maritime Affairs 25, DOI 10.1007/978-3-642-37033-5_3,

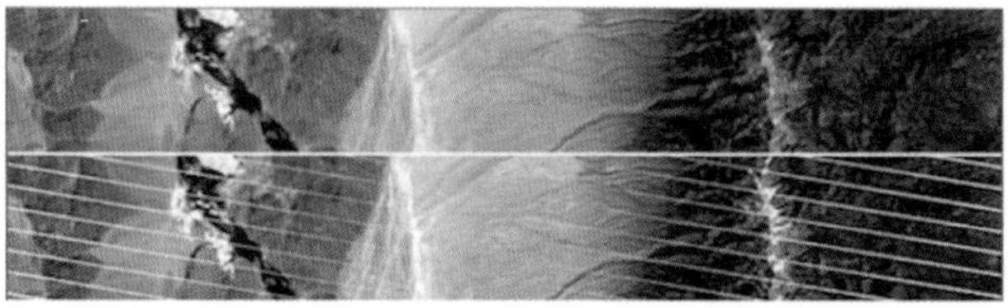

Fig. 3.1 *Top image*: pre-SLC anomaly scene. *Bottom image*: scene after SLC anomaly [38]

data gaps extending from the side edges to the middle of the image (see Fig. 3.1). This problem cannot be recovered and appears to be permanent. From this point in time Landsat 7 ETM+ is acquiring image data in the "SLC-off" mode. The SLC anomaly causes a data loss of approximately 22 % [9]. The gaps can only be filled by interpolation, but this procedure is not applicable for the determination of melt ponds.

3.1.1 Radiometric Calibration Procedure

The following equation is used to calibrate the level 1G DNs to absolute radiances R_λ in $W\,m^{-2}\,sr^{-1}\,nm^{-1}$ at the sensor's aperture

$$R_\lambda = \frac{Lmax_\lambda - Lmin_\lambda}{DN_{max} - DN_{min}} * (DN - DN_{min}) + Lmin_\lambda, \tag{3.1}$$

where $Lmin_\lambda$ and $Lmax_\lambda$ are the spectral radiances scaled to DN_{min}, respectively to DN_{max} for each band in $W\,m^{-2}\,sr^{-1}\,nm^{-1}$. These have to be extracted for each band gain (high or low) respectively. The gain values change as the detector looses responsivity and are issued as Calibration Parameter Files (CPF's) updated for distinctive image acquisition time frames. The CPF files can be obtained at http://landsat7.usgs.gov/cpf/cpf.php.

DN_{min} and DN_{max} represent the minimum and the maximum quantized calibrated pixel value. Here we use $DN_{min} = 1$ and $DN_{max} = 255$; DN is the quantized calibrated pixel value in DN.

After scaling DNs to absolute radiances for each wavelength (R_λ), these radiances are converted to planetary or top-of-the-atmosphere (TOA) reflectance values α_{TOA} using the following formula:

$$\alpha_{\mathrm{TOA}} = \frac{\pi R_\lambda d^2}{S_\lambda \cos\theta}, \tag{3.2}$$

where R_λ is the spectral radiance at the sensors aperture from Eq. (3.1), d is the distance Earth–sun in astronomical units, S_λ is the mean solar exoatmospheric irradiance in $\mathrm{W\,m^{-2}\,nm^{-1}}$, and $\cos\theta$ is the solar zenith angle in degrees (see Fig. 2.2 on page 10).

S_λ and d are found in the Landsat 7 Science Data Users Handbook.

3.1.2 Reprojection and Resampling

Landsat data are delivered as GeoTiff files in Universal Transverse Mercator (UTM) map projection. For a better display and comparison with different sea ice products, the data is reprojected onto a polar stereographic grid according to the specifications of the National Snow and Ice Data Center (NSIDC).[2] For the comparison with MODIS data, we resample the data to a common multiple grid size of 50 m × 50 m, using a nearest neighbor algorithm.

3.2 MODIS

The MODIS (Moderate Resolution Image Spectroradiometer) sensor is a key instrument on board the Terra and Aqua satellites. The satellites were launched on December 18, 1999 and May 4, 2002, respectively, in a sun-synchronous, low altitude (705 km), near-polar orbit and at an inclination of 98.21°. While Terra is on a descending node with a mean equator crossing time of 10:30 a.m., Aqua operates on an ascending node with an equatorial overpass at 1:30 p.m. The continuous data collection of the two sensors are continuing through today and providing a complete spatial coverage of the Arctic Ocean for the past 12 melt cycles.

MODIS is acquiring data in 36 spectral bands,[3] ranging from 0.4 to 14.4 μm, with a resolution from 250 m to 1 km. From MODIS observations, several data products to describe land, ocean, and atmosphere features are offered through the NASA Warehouse-Inventory-Search-Tool (WIST) website.[4] For this study, we select the MOD09 surface reflectance data set. This data set is based on the MODIS level 1B product. The level 1B product contains calibrated and geolocated at-aperture radiances in $W\,m^{-2}\,sr^{-1}\,nm^{-1}$ from all spectral bands of the MODIS sensor [48]. The MOD09 product is computed from the first seven bands of the level 1B product and is corrected from atmospheric scattering and absorption. Additionally, a BRDF correction should be applied (see also Sect. 2.2). In reality, however, the BRDF correction over sea ice is only performed in some coastal and shallow ocean regions.[5]

The MOD09 product is available in two levels: The level 2 product or MOD09GA is directly produced from the daily collected and to level 1B transformed MODIS data. Therefore, a global MOD09GA data set exists for each day. The best level 2 observations of the following 8-day period in respect of good viewing geometry, absence of clouds and cloud shadows, and aerosol loading are

---

[2] http://nsidc.org/data/polar_stereo/ps_grids.html, accessed in February 2012.

[3] Detailed wavelength ranges are listed in Appendix A.2.

[4] https://wist.echo.nasa.gov/api/, accessed in February 2012.

[5] Personal communication with Chrystal Schaaf, 24 August 2010.

used to produce the level 3 or MOD09A1 product. The resolution of both MOD09 products is 500 m × 500 m.

A detailed description of the product is given in the MODIS Surface Reflectance Users's Guide [82].

3.2.1 Surface Reflection and BRDF of Sea Ice

For the study on MODIS data, we act on the assumption that the spectral signature of MODIS' moderate resolution pixels always contain different surface types in the presence of sea ice. Therefore, we presume that the sea ice surface is comparable to the spectral signature of a Lambertian reflector (see Sect. 2.2) and a BRDF correction can be neglected.

To verify this assumption, available cloud-free MODIS level 1B data for a defined area and time period are selected. As test area, we chose Franklin Bay (69°40′N, 125°30′W). This bay is a large inlet in the Northwest Territories, Canada, located in the southeastern Beaufort Sea (see also Fig. 1.2 on page 4). The test period is June 2008. In the beginning of June, Franklin Bay is covered with fast ice and drift ice. The drift ice disappears in mid-June, the fast ice breaks off around June 23, so that the ocean is nearly ice free at the end of June (see Fig. 3.2). The selected subset contains open water areas, which represent the lowest TOA-reflectances in the MODIS scene with values around 0.2 in band 1 (459–475 nm). The snow covered sea ice areas represent the highest reflectances with values up to 0.8 in the same spectral range. Figures 3.3–3.5 describe the lowest and the highest values found in each data set as a function of the solar zenith angle, the relative solar azimuth angle, and the relative solar zenith angle, respectively. For the definitions of the angles, see Fig. 2.2 on page 10. We chose the lowest spectral range (band 1) for the analysis of the reflectances, because it is most sensitive to the high reflecting snowy surfaces (see also Fig. 2.7 on page 13). In Fig. 3.3, a dependence of the minimum values—representing open water—on the solar zenith is observable. Since a calm water surface is assumed to be a specular reflector, this dependence is expected. In contrast thereto, the maximum values—representing high reflecting surfaces—display no dependence on the solar zenith. The same effect can also be observed in Fig. 3.4 and 3.5. Since we cannot observe any dependence in the high values to the solar irradiance, the BRDF correction can be neglected for sea ice areas represented in MODIS pixels.

In a comparison of surface albedo of the Arctic Ocean from AVHRR data with SHEBA data it is described, that with the onset of sea ice melt, the surface becomes very inhomogeneous and is assumed to be a Lambertian reflector [86]. Additionally, after the statements from Grenfell (1977), the total albedo correction by using (i) empirical curves for the dependence of albedo on solar elevation and (ii) a specular correction based on the Fresnel reflection formulas especially for melt pond data is generally not more than 0.05 [26].

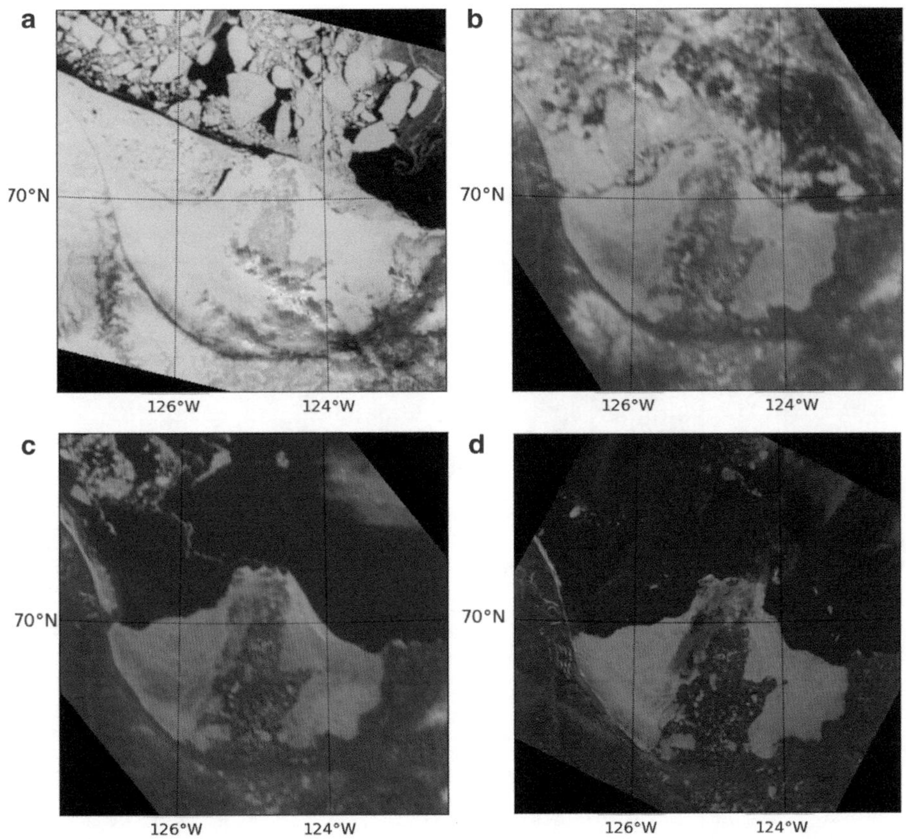

Fig. 3.2 Time series of MODIS images showing the Franklin Bay, Canada. The acquisition dates are (**a**) May 13, 2008, (**b**) June 3, 2008, (**c**) June 7, 2008, and (**d**) June 16, 2008. The images are true color composites (bands 2–4–3) of MODIS level 1B data

The both above given statements confirm the decision to neglect the BRDF correction for sea ice areas and to assume the sea ice surface as Lambertian reflector.

3.2.2 Data Masking

The MOD09 surface reflectance products used in this study are already corrected for atmospheric gases, aerosols, and thin cirrus cloud contamination.

To remove cloud pixel and land pixel, we use the cloud and land flags contained in the MOD09 product. From this flags both, a cloud mask and a land mask, are generated.

Clouds are generally characterized by higher reflectances and lower temperatures than the Earth's surface. This fact can be used to distinguish cloud pixels from clear sky pixels. The cloud masking algorithm is based on a multi-spectral thresholding

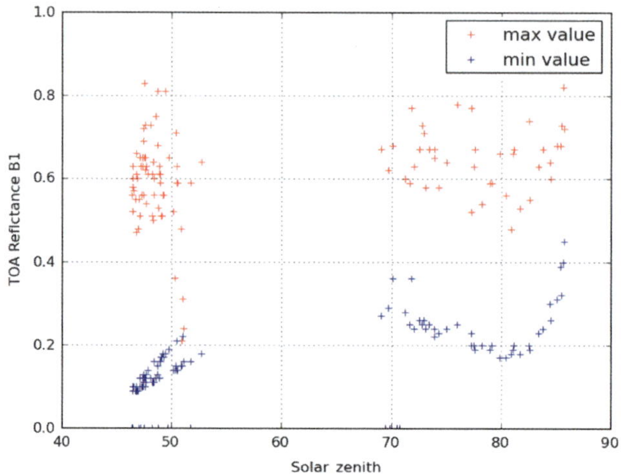

Fig. 3.3 Relation between solar zenith angles and TOA-reflectance values for level 1B data of band 1 (457–475 nm) from the MODIS sensor. Displayed are the highest value (max value) and the lowest value (min value) of each available data set of the Franklin Bay case study in June 2008

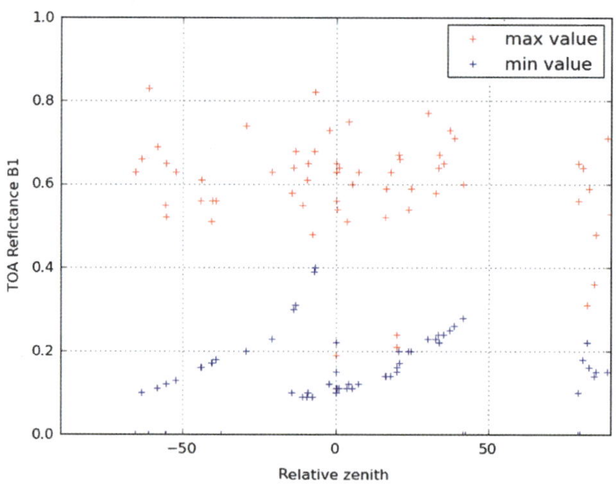

Fig. 3.4 Relation between relative zenith angles and TOA-reflectance values for level 1B data of band 1 (457–475 nm) from the MODIS sensor. Displayed are the highest value (max value) and the lowest value (min value) of each available data set of the Franklin Bay case study in June 2008

technique, applied to each pixel of the satellite overpass scene. A detailed description of the algorithms and thresholds used can be found in the Algorithm Theoretical Basis Document from the MODIS Cloud Mask Team [1]. From visual tests, we decide to define pixels with the attributes "cloudy" (confidence $<= 0.66$) and "probably cloudy" ($0.95 >=$ confidence > 0.66) as cloud-contaminated pixels

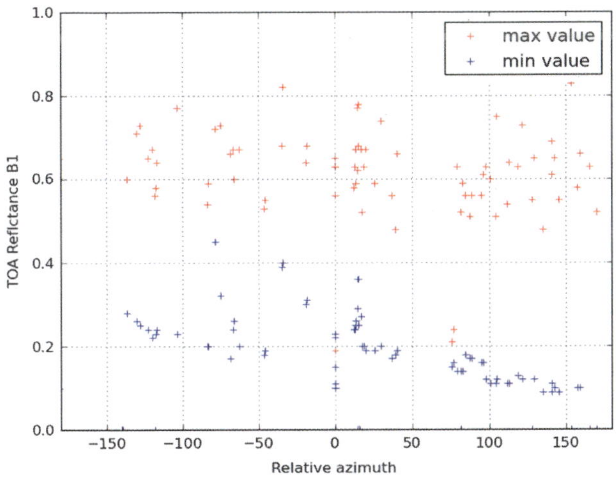

Fig. 3.5 Relation between relative azimuth angles and TOA-reflectance values for level 1B data of band 1 (457–475 nm) from the MODIS sensor. Displayed are the highest value (max value) and the lowest value (min value) of each available data set of the Franklin Bay case study in June 2008

and the attributes "probably clear" ($0.99 >= \text{confidence} > 0.95$) and "confident clear" ($\text{confidence} > 0.99$) as clear-sky pixels.

The MOD09 product additionally contains a land/water background flag. This flag defines four possible surface types: land, water, desert, coast. For our purposes the flags land, coast, and desert are merged into the land flag. The sum of all land flags yields a land mask.

3.2.3 Reprojection and Mosaicking

The MOD09 product is delivered as HDF[6] in a sinusoidal grid, containing 460 non-overlapping tiles. The original MODIS tiles are first reprojected onto a polar stereographic grid according to the specifications of the National Snow and Ice Data Center (NSIDC).[7] The single spectral bands are stored separately. For reprojection and processing GDAL (Geospatial Data Abstraction Library) [24] and Python SciPy [33] are used. The single spectral bands are merged to a GDAL data set, that contains all necessary information which are also included in the former HDF tile. From this GDAL data set the desired information are read out, e.g. spectral values of specific bands or flags. Cloud and land masks are created and overlaid over the single spectral bands (see Section above). Finally, the spectral bands are mosaicked

---

[6]Hierarchical Data Format.

[7]http://nsidc.org/data/polar_stereo/ps_grids.html, accessed in February 2012.

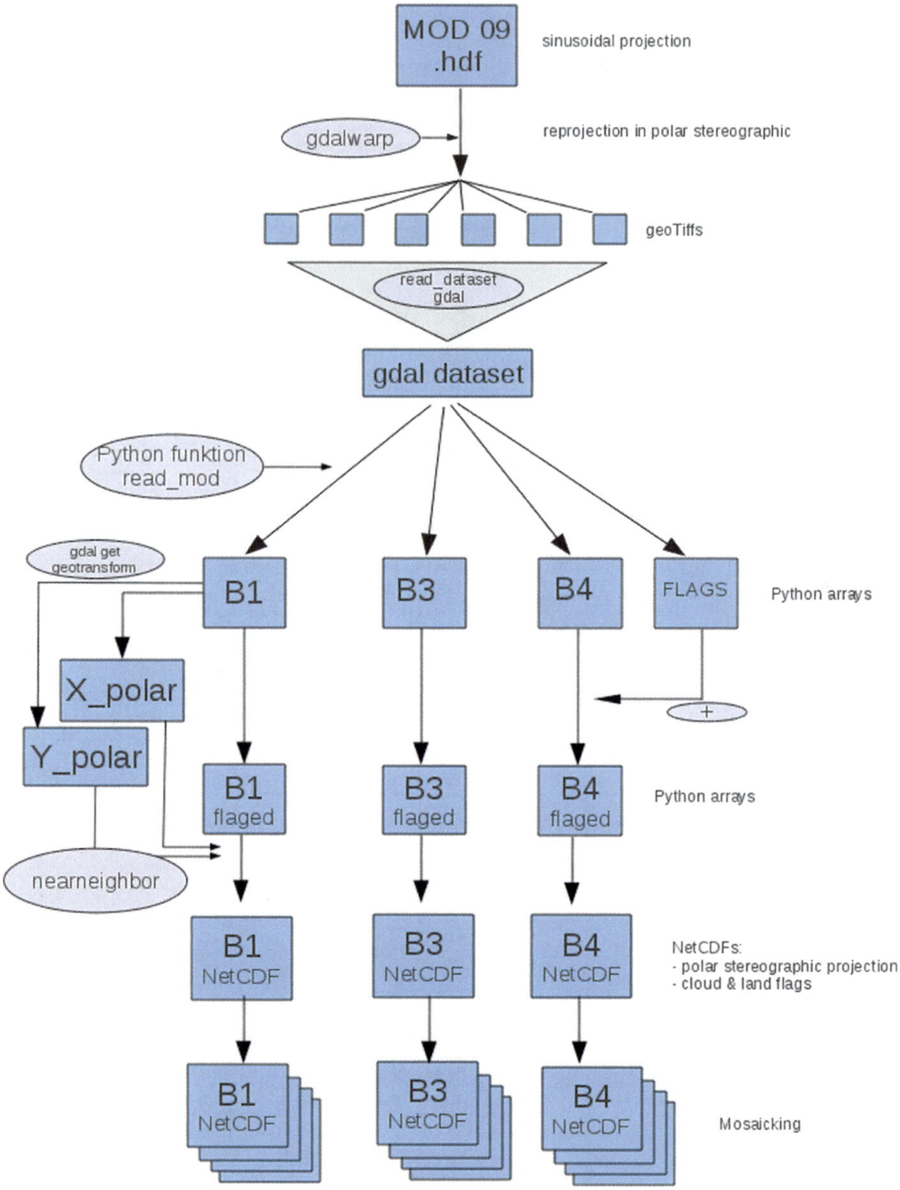

Fig. 3.6 Pre-processing steps for the MODIS surface reflectance product (MOD09)

band-wise and stored as NetCDF[8] (Fig. 3.6). These Arctic-wide mosaics are used as input data for further calculations.

---

[8]Network Common Data Form.

For the comparison with validation data or to gain a quick overview over an area, true color composites from the MODIS level 1B product, using the spectral band combination 2–4–3 are generated. The level 1B data are obtained from the website http://lance-modis.eosdis.nasa.gov/cgi-bin/imagery/realtime.cgi, accessed in February 2012.

Chapter 4
Melt Pond Determination from Landsat Satellite Data

4.1 Studies on Landsat Data

Since January 2009, Landsat satellite imagery is freely accessible through the archive of the US Geological Survey (USGS).[1] To use this potential, we conducted an extensive archive search with a subsequent case study. Due to strong limitations regarding coverage and pixel quality (see Sect. 4.1.1), the performed archive search yielded only two usable Landsat 7 ETM+ scenes over the Arctic sea ice area for the summer 2001. For the case study, we select an area of 25 km × 25 km in the northern Beaufort Sea at 80°–82°N, covered by a Landsat 7 ETM+ scene with a sun elevation angle of 28.7° which is acquired on July 18, 2001 (Fig. 4.1).

As can be seen from a true-color composite of the study area, shown in Fig. 4.2, the reflectivity of the different ice fractions varies significantly when using the Landsat band combination 3–2–1. Snow covered areas appear white. With the onset of the melting and wetting of snow and ice surfaces, the color spectrum varies across shades of gray and blue. Shallow melt ponds appear in light blue and change from a green–blue tone to a darker color once they have reached their full extend [43]. The color of a melt pond is not only depending on the depth but also on the texture of the underlaying surface [55].

Within this subset, reference areas for further analysis are defined. The reference areas or training areas cover all necessary features such as open water (A), snow covered ice (B), and two types of melt ponds (C and D). These features are visually identified from Fig. 4.2. We assume, that they are suitable to represent the specific classes.

[1]Landsat imagery held in the USGS archives can be searched on the following pages:

- Glovis: http://glovis.usgs.gov;
- Earth Explorer: http://earthexplorer.usgs.gov;

both accessed in February 2012.

A. Rösel, *Detection of Melt Ponds on Arctic Sea Ice with Optical Satellite Data*, Hamburg Studies on Maritime Affairs 25, DOI 10.1007/978-3-642-37033-5_4,

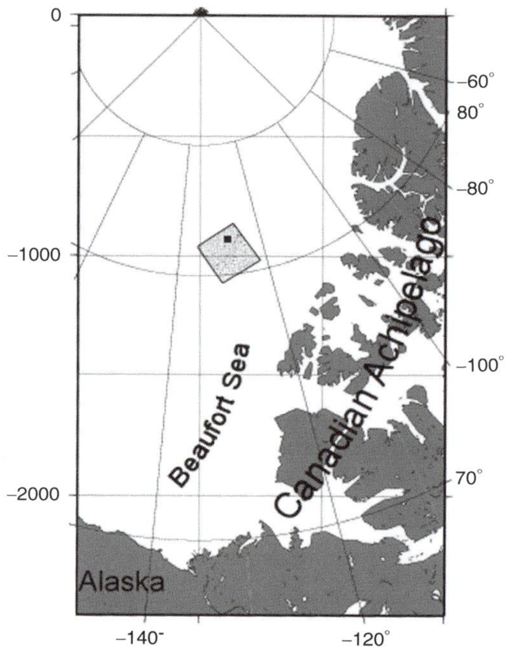

Fig. 4.1 Overview of the study area defining the location of the selected Landsat 7 scene from July 18, 2001 and the subset area (*black square*) used for this case study

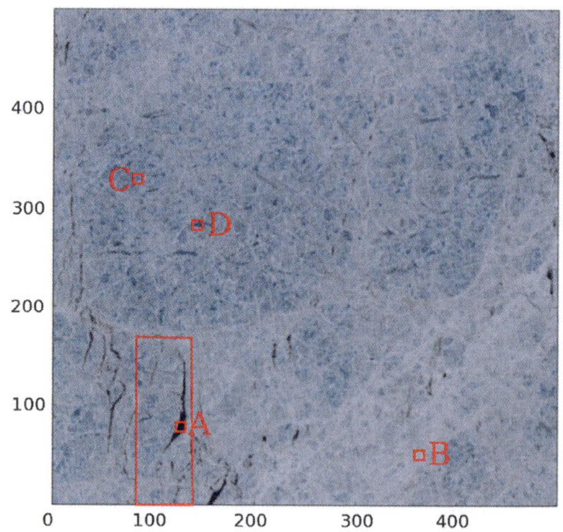

Fig. 4.2 25 km × 25 km true color subset (band combination 3–2–1) of the Landsat scene displayed in Fig. 4.1 showing the selected training areas (description in the text). This subset is used for all analysis. The *red box* shows the zoom in area for the comparison of the different methods (Fig. 4.7). The coordinates are the number of grid cells

4.1.1 Archive Search and Saturation Problems

The freely accessible Landsat archive with acquisitions from all Landsat sensors covering a time period from 1972 until today is a valuable data collection of Earth observation data, which appeared to be suitable to study melt ponds on the Arctic sea ice.

The aim was to use this extensive collection of satellite data from an optical sensor for determining melt ponds on the Arctic sea ice, therefore we started with an archive search of the USGS archive. During the archive search, we had to realize that most of the satellite scenes cover only coastal areas and scarcely open sea ice areas. The long revisiting time of 16 days and the high number of overcast days in the Arctic region are an additional limiting factor to the total number of suitable scenes. Moreover, the 8-bit Landsat sensors have severe problems with saturation over snow- and ice-covered surfaces which is also related to the solar elevation [4].

Saturated scenes are generally not applicable for the purpose of this study, since the methods used for analyzing the Landsat images are based on the spectral differences between the first four optical and NIR bands of the sensor. Subsequently, suitable scenes regarding location and cloud cover were downloaded to search for non-saturated data sets. The first four bands were read in and the maximum values of the DN were identified. If the maximum value is at least in one band equal to 255, the scene was flagged as saturated. This search was time-consuming and thus we aborted the search after we found two unsaturated scenes from July 18, 2001 (see Sect. 4.1) which are used in the further analysis as study area.

4.1.2 Determination of Melt Ponds After Markus et al. (2003)

The spectral differences of melt ponds in comparison to bare or snow covered sea ice and open water, as described in Sect. 2.3, are used to detect melt ponds with a method proposed by Markus et al. (2002, 2003) [43,44]. From Fig. 2.7 on page 13, it is evident that these spectral differences of the individual surface types increase with longer wavelengths.

The differences of two spectral bands of the Landsat 7 ETM+ sensor[2] are used in this method to detect melt ponds. The differences between band 2 and band 3 ($\Delta 23 = \alpha_{b2} - \alpha_{b3}$), which are located above a certain threshold, are classified as ponded area. Values below this threshold are categorized as open water and unponded ice areas. To distinguish also between open water and ice, the spectral information from band 1 (α_{b1}) is used: High values are classified as ice, low values as water. For the selected Landsat 7 subset $\Delta 23$ is plotted versus α_{b1} and the scatter plot is divided into three regions, using the following threshold and tie-points (see Fig. 4.3): $\Delta 23_T = 0.06$ distinguishes between melt ponds and open water/ice, additionally $\alpha_{b1T} = 0.16$ is defined as 100 % open water and $\alpha_{b1T} = 0.82$ as 100 % ice. If $\Delta 23 \geqq 0.06$ and $\alpha_{b1} = 0.45$, the pixel is defined as 100 % melt pond. The pixels of the selected training areas (A: open water, B: snow covered ice, C and D: two types of melt ponds, see Fig. 4.2) are highlighted in Fig. 4.3.

The described procedure strongly depends on the choice of the thresholds and tie-points. These values need to be determined separately for each satellite scene.

[2]Band specifications of the Landsat sensor is displayed in Appendix A.1.

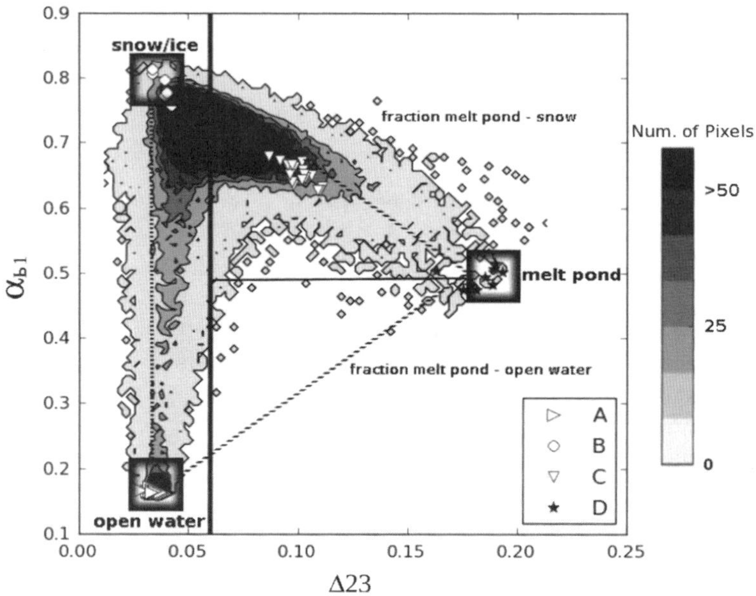

Fig. 4.3 Scheme of the tie point method for classifying Landsat images (after [44]). *A, B, C, D* show the pixels of the training areas for open water, snow covered ice and melt ponds (see Fig. 4.2)

Therefore, this method is not adequate for an automatic processing routine. We use the results of this method as a reference for comparison with a new method, introduced in Sect. 4.1.3.

4.1.3 Determination of Melt Ponds Based on Principal Component Analysis

To detect melt ponds with a method based on principal component analysis (PCA), the same spectral differences of melt ponds in comparison to bare or snow covered sea ice and open water as used in the melt pond determination after Markus et al. (2003) (see Sect. 4.1.2) are used.

For the analysis of the data using PCA, we decide to use the spectral bands that can best resolve these spectral differences. Figure 4.4 displays a two-dimensional (2-D) histogram plot of band 1 versus band 4 of the above described Landsat scene. Included are pixels of the determined training areas (see previous section). A maximum of pixels is detected at high reflectances in both bands. A second maximum is located at the low reflectances in both bands. Both maxima are represented through the training areas A (open water) and B (snow covered ice).

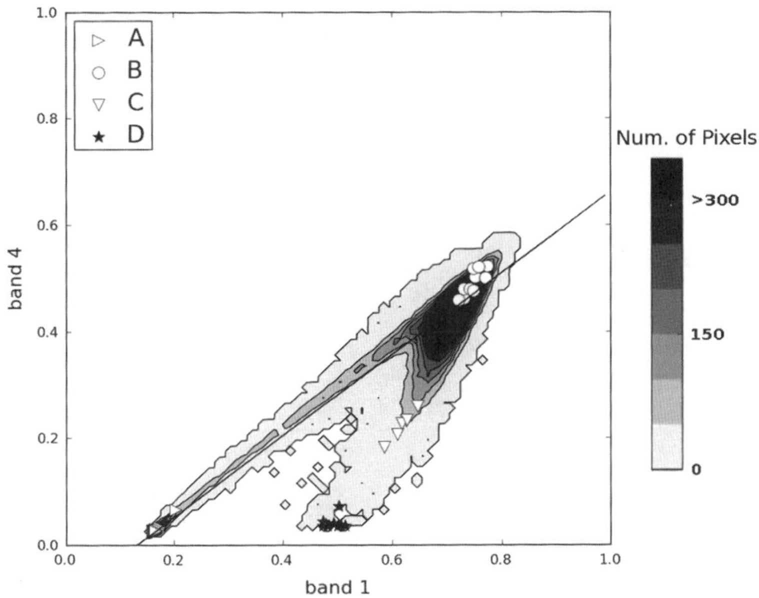

Fig. 4.4 2-D histogram of Landsat reflectivities of band 1 versus band 4 for the selected scene. *A, B, C, D* show the training areas for open water, snow covered ice and melt ponds (see Fig. 4.2). The *straight line* describes the mean vector [see Eq. (4.1)]

It is observable, that the main quantity of pixels—including classes A and B—lies on a main axis, with values that are highly correlated between the two bands; they represent the first principal component. Another cluster is found which includes those reflectances with lower values in band 4 than their pendants in band 1. This cluster describes pixels with an included melt pond fraction and represents the second principal component. Thus, the classes C and D can be found in this cluster.

To separate the pixels with a melt pond fraction from pixels which only have a snow and water signature, the PCA is applied. First, the mean vector of the first principal component ($g1$) and a rotation matrix must be defined. The second principal component ($g2$) is perpendicular to $g1$. Generally, $g1$ is evaluated from all pixels of the data set, but for this analysis, the mean vector is defined only for the "main axis" (open water–ice pixels) by using the two maxima of the 2-D histogram plot (see Fig. 4.4). The mean vector can be described with a linear equation:

$$f(x) = mx + b, \tag{4.1}$$

with slope m and offset b.

The rotation matrix is defined as:

$$R = \begin{bmatrix} \cos\gamma & -\sin\gamma \\ \sin\gamma & \cos\gamma \end{bmatrix}, \tag{4.2}$$

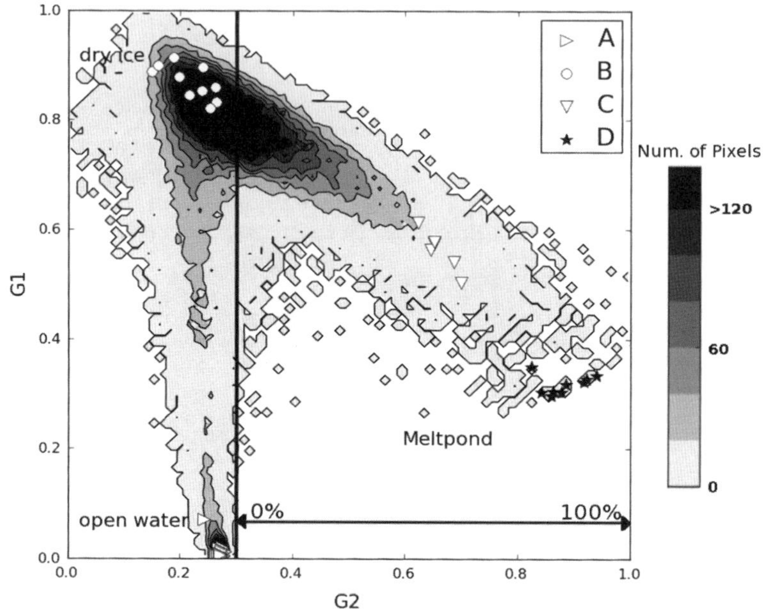

Fig. 4.5 Transformed, new coordinate system with $G2$ and $G1$. *A, B, C, D* show the training areas for open water, snow covered ice and two types of melt ponds (see Fig. 4.2)

where

$$\gamma = -\arctan m. \tag{4.3}$$

The individual components, also called eigenvectors, $g1$ and $g2$ describe the principal component axis of the original coordinate system. After the application of the transformation matrix, the pixels are positioned in a new coordinate system, where the eigenvectors $g1$ and $g2$ become the new axis $G1$ and $G2$ (see Fig. 4.5). Now, the data do not exhibit a discernible correlation between the "main axis" and the cluster of melt ponds. Furthermore, the transformed data set allows a distinction between the water–ice axis and the cluster of melt pond pixels.

A threshold to separate the water–ice pixels from the melt pond pixels is defined. This threshold runs parallel to the open water–ice axis and is identified as follows:

$$G1 > 0.2 \quad \text{and} \quad G2 = \max(G2). \tag{4.4}$$

For the analyzed Landsat subset, the threshold value is set to $G2_T = 0.3$. All pixels with values above the given threshold will be assigned as "melt pond pixels".

In the present case, melt pond areas are very well developed and it can be assumed that at least one pixel is covered completely with melt ponds. All other pixels are only fractionally ponded. The melt pond pixels are scaled between 0 % and 100 % (see Fig. 4.5).

To determine the melt pond fraction—that means only the fraction of melt ponds on the sea ice—the open water area needs to be defined and subtracted from the overall area. For determination of the open water area in Landsat data a threshold technique is applied, as already used by Steffen et al. (1991) and Cavalieri (2006) [8, 74]:

$$G1 < 0.2 \quad \text{and} \quad G2 < G2_T. \tag{4.5}$$

4.2 Results

4.2.1 *Sensitivities and Error Calculation*

The threshold, which divides pixels into ponded or unponded sea ice, is the input parameter in this analysis. The uncertainty of this method is mainly determined by the threshold value $G2 = 0.3$. The threshold in this procedure is depending on the solar zenith angle, the geographical latitude and subsequently on the spectral reflectance of the surface. Hence, the threshold must be defined for each satellite scene. Therefore, it is important to investigate how sensitive the results depend on a variation of $G2_T = \pm 1\,\%$ and $G2_T = \pm 5\,\%$, and correspondingly $\Delta 23_T = \pm 1\,\%$ and $\Delta 23_T = \pm 5\,\%$ for the method of Markus et al. (2003) [44].

For $G2_T = \pm 1\,\%$ the total melt pond area varies by 4.75 % and for $G2_T = \pm 5\,\%$ the area varies by 22.45 %. The almost linear increasing of the error in both procedures shows, that the distribution of the melt pond fraction is in this case a uniform distribution. The calculated errors for the PCA method in this case study are about 0.5 percentage points lower than for the method of Markus et al. (2003) [44].

For the error estimation, all mixed pixels at the open water/ice edge are marked and assigned as 50 % misclassification. This results in an error of ±0.35 % for the open water area for this subset.

Tests in the full Landsat scene show that a variation of ±5 % for the threshold is realistic. Hence, this value was used for the error calculation.

4.2.2 *Comparison of Two Methods*

The results from the classification method based on the PCA, as described in Sect. 4.1.3, is shown in Fig. 4.6b. The comparison of the melt pond fraction with the true-color composite (Fig. 4.6a) shows marked similarities.

Figure 4.6c displays the result of the classification with the method after Markus et al. (2003) [44] as described in Sect. 4.1.2. The melt pond fraction exhibits conformity with the results of the PCA and the true-color composite. The coefficient of determination for both methods is $R^2 = 0.94$.

Fig. 4.6 (**a**) 25 km × 25 km true color subset (band combination 3–2–1) from the Landsat scene displayed in Fig. 4.1 on page 28 with the selected training areas *A–D* (description in the text). This subset is used for all analysis. The *red box* shows the zoom in area for the comparison (Fig. 4.7). The coordinates are the number of the grid cells. (**b**) Melt pond fraction of the Landsat subset determined with PCA. *Black areas* are identified as open water. (**c**) Melt pond fraction of the Landsat subset determined with the method after Markus et al. (2003) [44]. *Black areas* are open water

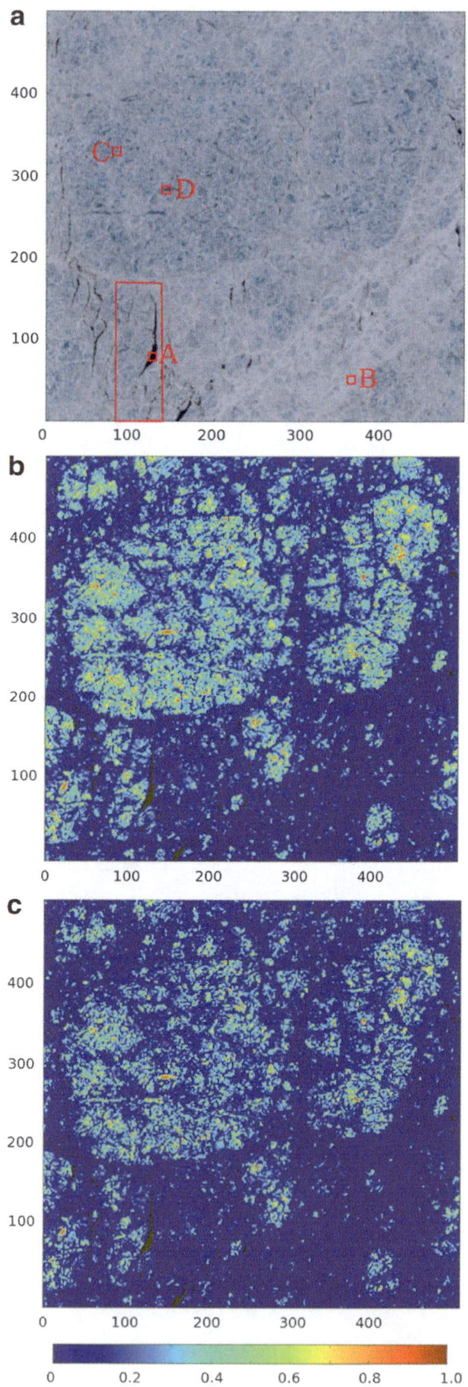

Table 4.1 Comparison of the results

	PCA	Markus
Melt pond coverage in km^2	106.3 ± 23.9	88.8 ± 10.6
Melt pond fraction in %	17.1 ± 3.8	14.3 ± 3.3

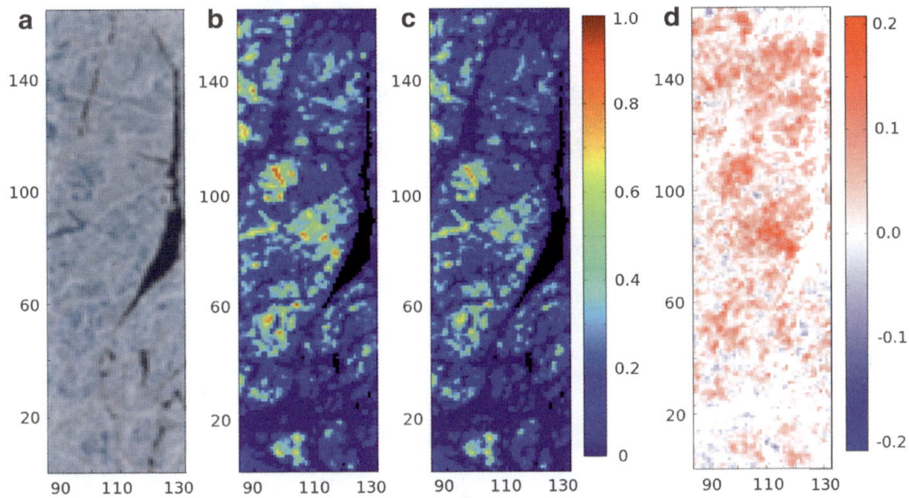

Fig. 4.7 Close up of the results shown in Fig. 4.6: (**a**) true color zoom in; (**b**) results of the PCA; (**c**) results of Markus' method; (**d**) differences of PCA and Markus' method

The results of both described methods are compared in Table 4.1: calculations performed with PCA result in a melt pond cover of 106.3 ± 23.9 km^2, which corresponds to a melt pond fraction of 17.1 ± 3.8 % of the sea ice area. After the method of Markus et al. (2003) [44] melt ponds cover 88.8 ± 10.6 km^2 of the total sea ice area, which results in a melt pond fraction of 14.3 ± 3.3 % of the sea ice area.

Figure 4.7d shows differences of the spatial melt pond distribution determined using PCA (Fig. 4.7b) and Markus' method (Fig. 4.7c), respectively. In areas with an apparently high melt pond fraction (>20 %), the results of the method after Markus and others (2003) give up to 20 % lower values than the PCA. On the contrary, in areas with a generally lower melt pond fraction (<20 %), this method gives up to 10 % higher values than the PCA.

4.3 Discussion and Conclusion

4.3.1 Discussion

Table 4.1 lists the results of the calculated melt pond fraction using the above described methods. Both methods using Landsat imagery are consistent.

The differences between both methods may result from the fact that three spectral bands (bands 1–2–3) are used in the Markus et al. (2003) [44] method. They cover only the short wavelength range of the spectral reflectances of the surface features. For the PCA method only band 1 and band 4 are used. Band 4 utilizes the advantage of larger spectral differences between melt ponds and ice or snow cover.

In addition, the method of Markus et al. (2003) [44] needs more manual input than the PCA method, because for each satellite scene tie-points for each fraction must be defined. For the PCA method, only the threshold that separates the melt pond fraction from the snow–water fraction needs to be defined.

Regarding the sensitivities of the thresholds, the lower values for the PCA method result from the sharp defined data separation into melt pond fraction and the snow–water fraction (see Fig. 4.5). It is recognizable from Fig. 4.3, that the ice–open water axis in the method after Markus is not exactly parallel to the threshold. This effect causes the slightly more diffuse intersection between the two fractions.

4.3.2 Conclusion

Very high resolution satellite data are potentially useful to determine melting features on Arctic sea ice. This can be performed by analyzing the spectral differences of different surface types.

However, an extensive archive search of the USGS archive for suitable Landsat scenes yielded only a small number of usable scenes. A closer inspection of the data reveals that nearly all scenes, which potentially could contain melt ponds, have problems caused by saturated pixels. Saturation is related to a high sun elevation and a highly reflecting surface type. Existing procedures to interpolate between the saturated and unsaturated bands could not easily be applied, since the different spectral signals of the surface types have to be utilized. Another disadvantage is the revisit time of the ETM+ sensor every 16 days. Additionally, the Arctic is an area with an above-average amount of cloudy days, which limit the number of usable satellite scenes.

The above mentioned facts show clearly, that very high resolution satellite data can additionally be used for detecting melt ponds and can therefore provide a basis for comparison with lower resolution satellite data, but they cannot be used as the only data source for the creation of a seasonal melt pond data set of the entire Arctic.

Chapter 5
Melt Pond Determination from MODIS Data

As demonstrated in Sect. 4.1, the extensive Landsat archive is not suitable to generate neither a multi-annual nor an Arctic-wide melt pond data set. Because of the comparatively lower resolution and wider swath width of the MODIS sensor, a continuous spatial coverage of the Arctic with satellite data is available. There is however, a downside to the lower resolution. Given, that one MODIS pixel represents an area of 250 m × 250 m, it must be assumed that each pixel above sea ice contains more than one surface type. In remote sensing, these pixels are called "mixed pixels". Since no specific surface type can be assigned to the "mixed pixels", the two in Chap. 4 described methods for determining melt ponds from Landsat data are not applicable on MODIS data. Therefore, another approach which considers these "mixed pixels" is introduced.

5.1 Method

As already described in Sect. 2.3, the spectral albedo of sea ice is a function of the wavelength of incident solar radiation (see Fig. 2.7 on page 13). Highest albedo values appear for dry snow in the short wavelength range from 400 to 600 nm. The albedo is decreasing towards longer wavelengths. At 500 nm, melt ponds have albedo values between 0.6 for young and shallow ponds and 0.25 for mature ponds. As mentioned before, albedo values for melt ponds depend on the depth and underlying surface of the ponds.

Snow covered ice, bare ice, wet snow, and melting bare ice show a smaller reduction in spectral albedo at higher wavelengths than the different types of melt ponds. The albedo of ponded ice is characterized by a decrease between 500 and 800 nm. In this wavelength range the absorption of water becomes a dominant factor and is responsible for the reduction of the spectral albedo. Above 800 nm, the radiation is nearly totally absorbed by the water surface, and the underlying ice layer has no influence on the albedo [26]. The spectral curve for open water (Fig. 2.7h on page 13) shows values around 0.06 through the entire wavelength spectrum.

A. Rösel, *Detection of Melt Ponds on Arctic Sea Ice with Optical Satellite Data*, Hamburg Studies on Maritime Affairs 25, DOI 10.1007/978-3-642-37033-5_5,

Table 5.1 Spectral reflectances (r_i) of surface types used in the unmixing algorithm [81]

MODIS band	Bandwidth (nm)	Resolution (m)	Melt pond r_M	Snow/ice r_I	Open water r_W
1	459–479	500	0.22	0.95	0.08
3	620–670	250	0.16	0.95	0.08
4	841–876	250	0.07	0.87	0.08

5.1.1 Three-Surface Class Model

Based on the differences of the spectral curves in Fig. 2.7 on page 13, we decide to distinguish between three surface types: open water (W), melt ponds (M), and snow and ice (I) [63, 81]. To calculate the albedo of sea ice on a large scale, the surface-based albedo values are weighted with the fraction of their corresponding surface component [21, 55, 81].

The so-called areal-averaged albedo $\overline{\alpha}$ for larger areas containing three surface types is then expressed as

$$\overline{\alpha} = \alpha_W A_W + \alpha_M A_M + \alpha_I A_I, \quad (5.1)$$

where A_i is the area fraction and α_i is the wavelength-integrated albedo.

The identification of the three surface types on Arctic sea ice is based upon a spectral unmixing procedure of multi-spectral satellite images and proposed by Tschudi et al. (2005, 2008) [80, 81].

It is based on the following equations:

$$\begin{aligned} A_W r_W(\lambda_1) + A_M r_M(\lambda_1) + A_I r_I(\lambda_1) &= R(\lambda_1) \\ A_W r_W(\lambda_3) + A_M r_M(\lambda_3) + A_I r_I(\lambda_3) &= R(\lambda_3) \\ A_W r_W(\lambda_4) + A_M r_M(\lambda_4) + A_I r_I(\lambda_4) &= R(\lambda_4) \\ A_W + A_M + A_I &= 1, \end{aligned} \quad (5.2)$$

where $R(\lambda_k)$ is the reflectance of each band $k = 1, 3$, and 4, with the corresponding wavelengths $\lambda_1 = 459$–479 nm, $\lambda_3 = 620$–670 nm, and $\lambda_4 = 841$–876 nm, for each MODIS pixel. A_i is the fractional coverage of each surface type, and $r_i(\lambda_k)$ represents the spectral reflectance for each surface type and for each band. The specific reflectance values for the three surface types used for these equations are listed in Table 5.1.

The system of linear equations (5.2) contains three unknowns (A_W, A_M, A_I) in four equations, therefore this system of equations has to be considered as overdetermined. That means (i) more than one exact solution can be possible, (ii) there is no solution, and (iii) there is an infinite number of solutions. Thus, we treat the linear equations (5.2) as an optimization problem. For the solution of these equation we use a quasi-Newton approximation method (see Sect. 5.1.2).

With the assumption of a three class mixture model and the selection of three surface types, we find that especially equations describing the surface types open water and melt ponds are almost linearly dependent. Therefore the system of linear equations (5.2) is not well conditioned. To comply with physical principles, it is necessary to constrain the solution interval between zero and one for each class. This side condition reduces additionally the linear dependency and is expressed by a sigmoid function,[1] implemented in a cost function $\mathbf{F_{cost}}$:

$$\mathbf{F_{cost}} = [(\mathbf{r} \cdot \mathbf{x}) - \mathbf{R}] + [(1 - \tanh(\mathbf{x}\gamma) - \tanh((\mathbf{x} - 1)\gamma)]w, \tag{5.3}$$

where $\mathbf{r}$ is the vector of the spectral reflectance values for each surface type, $\mathbf{x}$ is a vector of initial guess values for the fractions of surface types, $\mathbf{R}$ is the vector of the measured reflectance values ($\mathbf{R} = [R(\lambda_1), R(\lambda_3), R(\lambda_4)]^T$), γ is a gradient to define the slope of the function at the values zero and one, and w is a weighting factor. We chose as initial guess values $\mathbf{x} = [0.25, 0.25, 0.25]^T$, as gradient $\gamma = 150$, and as weighting factor $w = 0.1$. However, the results do not depend on the choice of $\mathbf{x}$ and w. This indicates, that the method is robust and the result is unique. On the one hand, the implementation of the cost function constrains the range of solutions and enhances the condition of the equation system, but on the other hand, this limitation causes higher computational costs compared to the solution of the linear equations (5.2) alone.

The solution of Eq. (5.3) provides the surface fractions of melt ponds, open water, and snow and ice. From this output the sea ice concentration A_{C} is quantified as:

$$A_{\mathrm{C}} = 1 - A_{\mathrm{W}}. \tag{5.4}$$

The melt pond fraction is defined as ponded area relative to the sea ice surface. Therefore, it is necessary to scale the retrieved melt pond fraction with the sea ice concentration. We define the relative sea ice melt pond fraction as

$$\widetilde{A_{\mathrm{M}}} = A_{\mathrm{M}}/A_{\mathrm{C}}. \tag{5.5}$$

For further analysis, the relative melt pond fraction and the sea ice concentration are gridded to a 12.5 km grid. Additionally, we calculate the amount of pixel used for creating the mean value in the 12.5 km-gridding routine (Fig. 5.1). This product can be used for further analysis to mask the melt pond fraction on a 12.5 km grid and it can also be considered as an indicator how trustworthy the result of the coarse grid is. A large amount of valid observations indicates high data quality; whereas, a low amount of valid observations indicates low data quality. To ensure high quality of the melt pond results, we create a mask of 12.5 km grid cells that contains at least 50 % cloud-free pixels. In the following context, this product is called "pixel-weight-mask based on a 50 % threshold". Reasons for this masking procedure are described in Sect. 5.2.3.

[1] The used approximation method cannot handle sharp edges of a rectangular function.

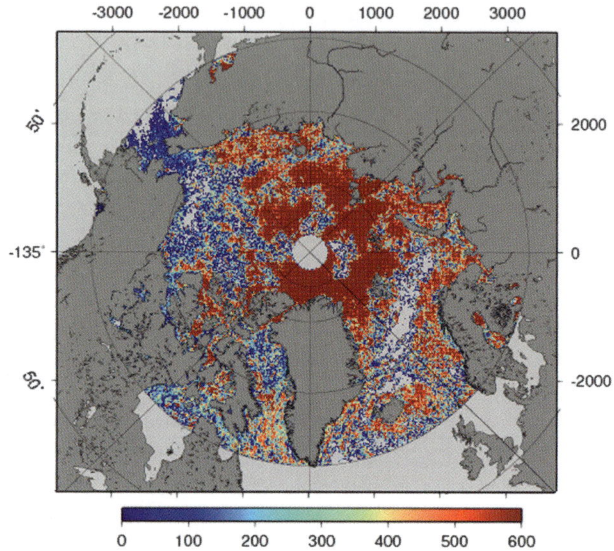

Fig. 5.1 Number of pixels involved into the 12.5 km grid. This example is from the data set from July 23, 2003. *Red values* indicate a high data density, *blue values* indicate a low data density, *gray areas* indicate data gaps and land mask

5.1.2 Regularization

As described in Sect. 5.1.1, Eq. (5.3) is both, overdetermined and not well conditioned. Especially, the reflectance curves for open water and melt ponds are linearly dependent. Thus, we assume Eq. (5.3) as optimization problem. Overdetermined linear equation systems can be solved with approximation methods. In numerical optimization, the Broyden–Fletcher–Goldfarb–Shanno (BFGS) algorithm is a method for solving optimization problems. The BFGS method approximates Newton's method[2] and is described as a hill-climbing optimization technique that seeks a stationary point of a function [51]. This routine uses the gradient of the objective function. If the gradient is not given by the user, then it is estimated using an initial guess. We compare the results of the calculations solved with the BFGS method and with the Nelder–Mead Simplex algorithm (fmin algorithm) regarding accuracy and computing time. As test area we chose a data set from June 2008 of Franklin Bay and create two subsets with 50×100 pixel, and $500 \times 1{,}000$ pixel, respectively (see Fig. 5.2).

[2]Newton's method assumes that the function can be locally approximated as a quadratic in the region around the optimum, and use the first and second derivatives to find the stationary point.

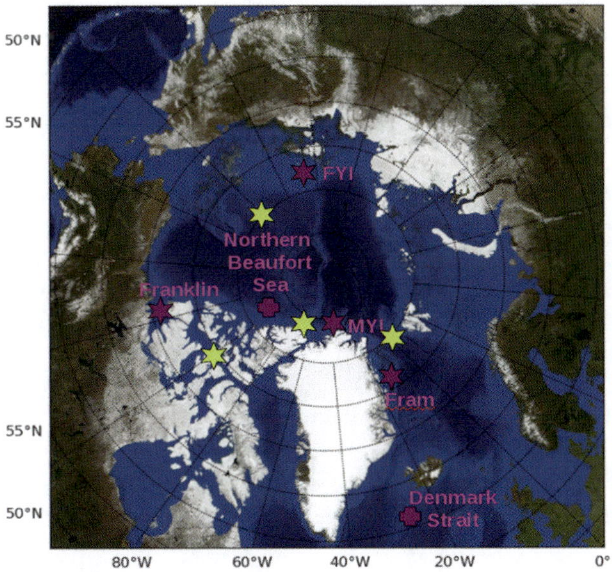

Fig. 5.2 Training and test sites for the used optimization method and ANN. Franklin Bay (*purple star*) is used for the optimization test (Sect. 5.1.2). Franklin, FYI, MYI, and Fram (*purple stars*) are training data sites for the ANN (Sect. 5.1.3). Northern Beaufort Sea and Denmark Strait (*purple cross*) are test sites for the systematic error of the ANN (Sect. 5.2.1). The *yellow stars* indicate four sites for the independent accuracy tests of the ANN (Sect. 5.2.1)

Table 5.2 Comparison of two optimization methods: fmin and BFGS

Test area	Method	Additional parameter	Melt pond fraction	Duration (s)
50 rows, 100 columns	BFGS	gtol = 1.0e^{-5}	0.1052	10.4
50 rows, 100 columns	BFGS	gtol = 0.1	0.0990	7.0
50 rows, 100 columns	BFGS	gtol = 0.2	0.1339	5.5
50 rows, 100 columns	BFGS	gtol = 0.5	0.1601	4.1
50 rows, 100 columns	fmin	xtol = 1.0e^{-5}	0.1052	21.0
50 rows, 100 columns	fmin	xtol = 0.1	0.1172	11.9
500 rows, 1,000 columns	BFGS	gtol = 1.0e^{-5}	0.0356	635
500 rows, 1,000 columns	BFGS	gtol = 0.1	0.0415	498
500 rows, 1,000 columns	fmin	xtol = 1.0e^{-5}	0.0356	1,008
500 rows, 1,000 columns	fmin	xtol = 0.1	0.0678	567

Using a strict tolerance value for the error (1.0e^{-5}), both test results show no differences in the resulting melt pond fraction, but the time reduction for the BFGS algorithm is around 40 % when looking at the larger area (see Table 5.2).

5.1.3 Artificial Neural Networks

Although we chose the faster BFGS algorithm for solving the set of equations, the average calculation time for a full MODIS scene is between 5 and 6 h.[3]

In order to process large areas or time series, this procedure is thus not applicable. To speed up processing, we use an artificial neural network (ANN).

The use of ANN for remote sensing data has been motivated by the realization that the human brain is very efficient at processing a huge amount of information in a very short time. Neurons in the human brain receive inputs from other neurons and produce an output which is then passed to other neurons. This cross-linking is the basis for the ANN architecture.

A multilayer perceptron (MLP) consists of multiple layers of computational units (= nodes or neurons), usually interconnected in a feed-forward way. That means, each node in one layer has directed connections to the nodes of the next adjacent layer, but has no connections to the nodes of the previous layers. As the signal passes from node to node, it is modified by weights $(w_1, \ldots, w_n)$ associated with the connections [3]. These weight values are established by a supervised learning technique, using a priory information about the actual output corresponding to the input data [25]. A typical MLP structure includes an input layer, one or more hidden layers and an output layer. The input layer only distributes the input signal into the network to the nodes of the first hidden layer without processing them. The nodes in the hidden layers and the output layer transform their input signal usually using a non-linear sigmoid function [3].

With the open source package FFNET for Python [85] we build a feed-forward MLP, composed of an input layer, two hidden layers with 9 neurons in the first and 27 neurons in the second layer, and one output layer (see Fig. 5.3). In this case, the number of neurons of the input layer is equal to the number of neurons of the output layer. For our purposes, the MLP is trained with back-propagation as learning technique [3, 66] using 5,000 learning steps. The number of layers, neurons and learning steps are determined by following the trial-and-error approach.

To train the ANN, an existing data set of melt ponds needs to be provided as training data set: This training data set is yielded from a limited number of pixels of different surface types and different dates by solving equation (5.3). Dates and amount of pixels of the chosen training data are listed in Table 5.3, the sites are marked with purple stars in Fig. 5.2.

The surface fractions achieved from the training data, namely open water, snow and ice, and melt ponds are subsequently used as input data to train the ANN.

---

[3]Calculations were performed on CliSAP's Linux servers snow with hardware specifications as follows:

- Processors: 8 x Quad-Core AMD Opteron 8356, 2.3 GHz (32 cores in total)
- Memory: 128 GB RAM

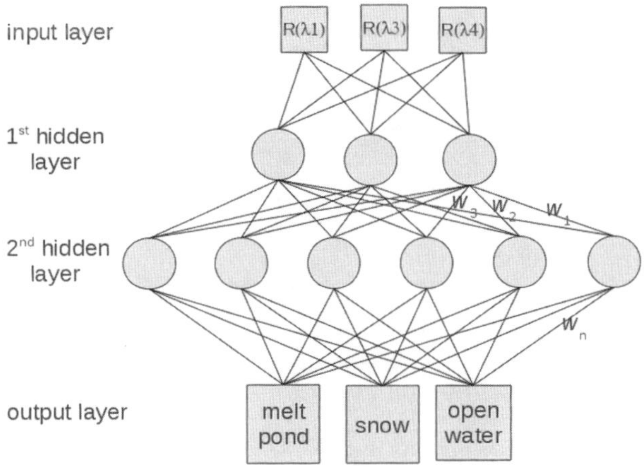

Fig. 5.3 Architecture of the trained MLP network used for this study. The input layer consists of MODIS surface reflectances for the three bands 1, 3, and 4 [see Eq. (5.2)]. The hidden layers of the MLP contains 9 knots in the first layer and 27 knots in the second layer. For the sake of clarity this figure shows only three knots for the first and six knots for the second layer. The connections between the single knots are associated with weights ($w_1, w_2, w_3, \ldots w_n$), determined from the back-propagation learning steps. The output layer contains three classes: melt pond, snow and ice, and open water

Table 5.3 Calculated melt pond, snow and ice, as well as open water fractions of ten training sites are used as input data to train the ANN

Data sets	Site name	Number of pixel	Surface types
May 8, 2008	Fram	1,200	Fast ice, multi-year ice
May 8, 2008	Franklin	400	Fast ice, first-year ice
May 8, 2008	FYI	400	First-year ice
June 9, 2008	Fram	1,200	Fast ice, open water, multi-year ice, floes
June 9, 2008	Franklin	400	Fast ice, open water, floes
June 9, 2008	FYI	400	First-year ice, open water
June 9, 2008	MYI	1,600	Multi-year ice
July 19, 2008	Fram	1200	Fast ice, open water, multi-year ice, floes
July 19, 2008	MYI	1600	Multi-year ice
September 5, 2008	MYI	1600	multi-year ice

To asses the consistence and the accuracy of the ANN, we perform two tests: Firstly, the training data set is used as test data set. Secondly, independent test data sets are selected. The results of the performance regarding consistence and accuracy of the ANN are presented in Sect. 5.2.1.

The fractional coverage of water, ice, and melt ponds is derived for the entire Arctic by using the trained ANN and MOD09 surface reflectance mosaics. Subsequently, the sea ice concentration and the relative melt pond fraction is calculated using Eqs. (5.4) and (5.5), respectively. These results are described in Sect. 5.2.3.

5.2 Results

In the following section we present the results of the study on MODIS surface reflectance data. Starting with a description of the accuracy of the artificial neural network (ANN) as well as a presentation of three validation studies to ensure the quality of our results, we then introduce both retrieved data sets: the MODIS sea ice concentrations and the MODIS melt pond fractions. Thereafter follows a Section that displays the results of a time series analysis of MODIS melt pond fractions, focusing mainly on the years 2007 and 2011. We conclude with a case study of the influence of melt ponds on microwave sea ice concentration retrieval methods. This case study illustrates the high potential of the melt pond data set for further applications.

5.2.1 Accuracy of the ANN

To evaluate the consistence of the ANN, we first apply the trained ANN on the selected training data sets (see Fig. 5.2) and compare the results with those calculated using the unmixing algorithm with the implemented side condition [see Eq. (5.3)]. The difference between the two results for this case is +0.1 %.

Thereafter, we operate the ANN on eight independent (regarding date and region) test data sets. These test data sets are from four different sites in the Arctic Ocean (see Fig. 5.2) from day 129, 161, 201, and 249 of the year 2010. Again, we compare the ANN-results of the test data sets with the results determined using Eq. (5.3). Here, the difference is with 0.8 ± 0.5 % only slightly higher than the previous test. The results of these two tests indicate, that the training of the ANN was successfully performed.

For an estimation of the systematic error of this procedure, we perform two further tests based on the following assumptions:

- An area of 50 km $\times$ 50 km around the coordinates 61.5°N and 26.0°W (Denmark Strait, see Fig. 5.2) is an open water area in June 2008 (day 161).
- Melting has not started on day 129 of the year 2008 at a latitude of 82.5°N (Markus et al., 2009) [45]. An area of 200 km $\times$ 100 km around the coordinates 82.5°N and 92.5°W (Northern Beaufort Sea, see Fig. 5.2) is covered with dry snow and/or bare cold ice in the first week of May 2008 (day 129). No melt ponds are observable.

The test for the open water area results in a relative error of +0.1 % for the estimated melt pond fraction, for the test on the snowy surface the relative error is +6 %.

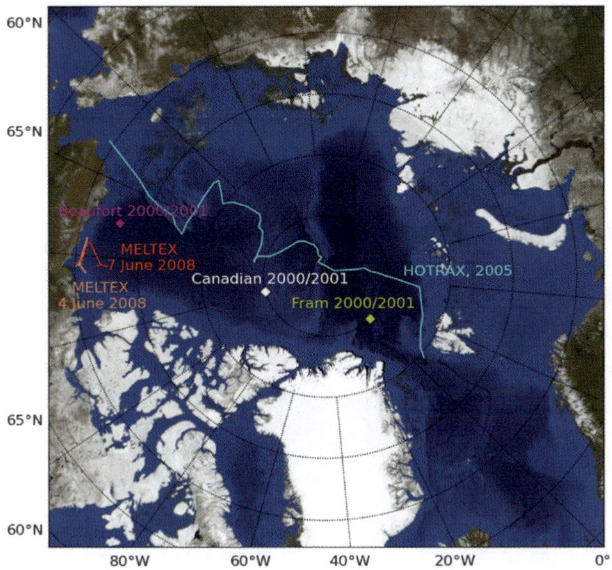

Fig. 5.4 Tracks, locations and dates of validation data sets. Detailed descriptions are given in the text (Sect. 5.2.2)

5.2.2 Validation

For validation, we use three different data sets: (i) the results of the aircraft campaign MELTEX ("Impact of melt ponds on energy and momentum fluxes between atmosphere and sea ice") conducted by the Alfred Wegener Institute for Polar and Marine Research (AWI) in May and June 2008 over the Beaufort Sea [5]; (ii) the melt pond data set from the National Snow and Ice Data Center (NSIDC) [22], and (iii) the results of the melt pond observations from the HOTRAX 2005 cruise [55] (Fig. 5.4).

5.2.2.1 Validation with MELTEX Data

The MELTEX campaign aimed to improve the quantitative understanding of the impact of melt ponds on radiation, heat, and momentum fluxes over Arctic sea ice. For this purpose, the BASLER BT-67 type aircraft POLAR 5 was employed, which had a downward-looking digital photo camera aboard. Weather conditions in May were mainly characterized by cold-air advection from the inner parts of the Arctic towards the coast of the southern Beaufort Sea. These cold air flows caused a refreezing of most melt ponds, which were still very shallow at that time. Even a thin layer of fresh snow on the refrozen ponds was observed. During the last week of measurements, in the beginning of June, a tongue of very warm air was shifted from

Alaska to Beaufort Sea. It reached its largest extension over the ocean on June 4 and June 5, 2008, which strongly forced the development of melt ponds.

Aerial photography was carried out with a digital reflex camera and a lens with an angle of view of 114°. A photo was taken every 6 s. In the preprocessing, all photos with sunglint were eliminated, and a cross-track illumination correction was carried out. Subsequently, a supervised classification employing the maximum likelihood method was performed to derive the areal fraction of the five surface classes distinguished: open water, thin ice, bare ice, snow, and melt ponds. Training data was manually defined in the photos. The probability threshold for the classification was set to 0.95. Pixels, which fell below this threshold were not classified. Only images with a number of unclassified pixels below 5 % were included in the present study. Because of the very shallow and partly refrozen ponds, the determination of the melt pond fraction based on an analysis of the aerial photos is difficult for days previous to June 4, 2008. In the present study, we therefore use only photos from two flights performed on June 4, and June 7, 2008 for the validation. The surface area covered by the photos is determined by the flight level of the aircraft. At 1,000 m, a photo covers an area of 1.4 km × 2.0 km, whereas at 30 m, a photo covers an area of 40 m × 60 m. For further comparison, the melt pond set is filtered by a Gaussian filter with $\sigma = 10$ to smooth the function, and to to avoid oversampling, only every tenth value is used. The classification of the aerial photos was performed at the Alfred-Wegener-Institute.

For the validation, we compare the MELTEX melt pond data with MODIS melt pond fractions for June 4, and June 7, 2008 (Figs. 5.5 and 5.6). In the true color composite (Figs. 5.5a and 5.6a), melt ponds can be determined as bluish areas on the individual ice floes, with shades ranging from light blue to dark violet.

By following the MELTEX flight track in both figures, the changing from medium melt pond fractions (20–25 %) to high melt pond fractions (28 %) is reflected in the underlying MODIS melt pond fractions. The transition to the open water areas is also characterized by a decreasing MODIS melt pond fraction.

The flight on June 4 started only at 18:03 UTC and ended at 00:09 UTC the next day. After the comparison of the daily level 2 MODIS data from June 4 with the individual level 1B acquisitions, we conclude that the corresponding area in the MOD09GA product originates from an acquisition in the early morning (4:00 UTC or 6:15 UTC). The same is valid for the MOD09GA product of June 5. Because of the rapidly changing meteorological conditions on June 4 in combination with the late flight time (18:03 UTC until 00:09 UTC) and the early acquisitions of the MODIS scenes, we use the MODIS melt ponds determined from the scene from June 5 for the validation. To fill the gaps in the MODIS melt pond sets caused by cloud filtering, we select the corresponding pixels from the melt pond sets from the days before and after the relevant data (Figs. 5.5b and 5.6b).

The average MODIS melt pond fractions for the entire flight tracks are 28.4 ± 2.5 % and 21.6 ± 8.1 % for June 4 and June 7. The corresponding MELTEX melt pond fractions are slightly lower (26.4 ± 11.5 % and 19.3 ± 13.5 %). The calculated root mean square errors (RMSE) for the dates June 4 and June 7 are 11.2 % and 10.6 % respectively.

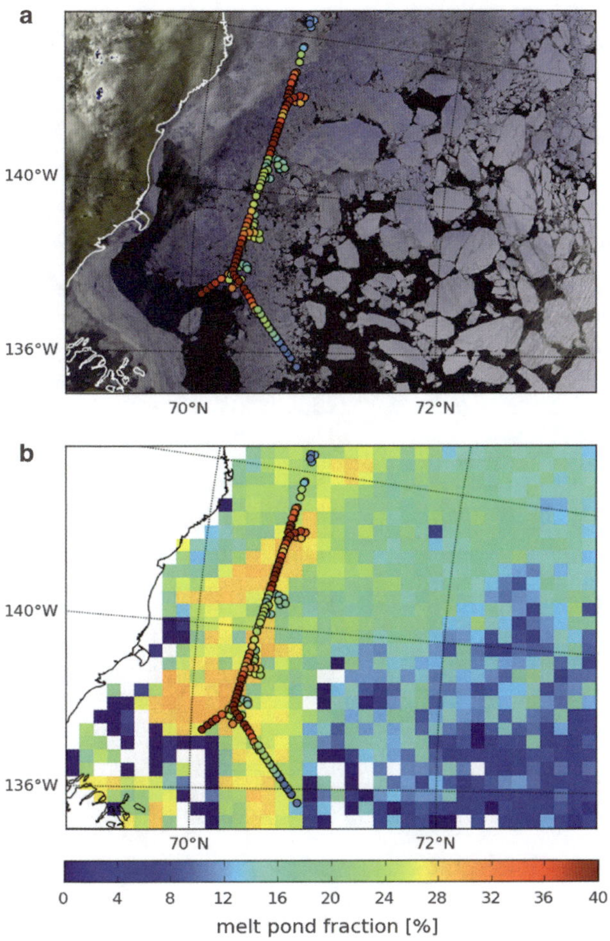

Fig. 5.5 (**a**) Spatial overlay of MELTEX melt pond results (*dots*) on a true color composite of MODIS level 1B data from June 4, 2008, 22:15 UTC. (**b**) MELTEX melt pond results (*dots*) from June 4, 2008 overlaid on MODIS melt pond fractions from mainly June 5, 2008 (detailed description in the text, Sect. 5.2.2.1)

5.2.2.2 Alidation Data from National Snow and Ice Data Center (NSIDC)

Sea ice melt pond statistics and maps over four Arctic ocean sites during the summers of 1999, 2000, and 2001, derived from visible band imagery from high resolution intelligence satellite data with a spatial resolution of 1 m, are mapped from 10 km × 10 km satellite scenes [22, 36]. These data sets were analyzed using a supervised maximum likelihood classification to derive either two (water and ice) or three (water, ice, and melt ponds) surface classes [22]. The resulting melt pond data set consists of tables of pond coverage and size statistics for 500 m × 500 m cells.

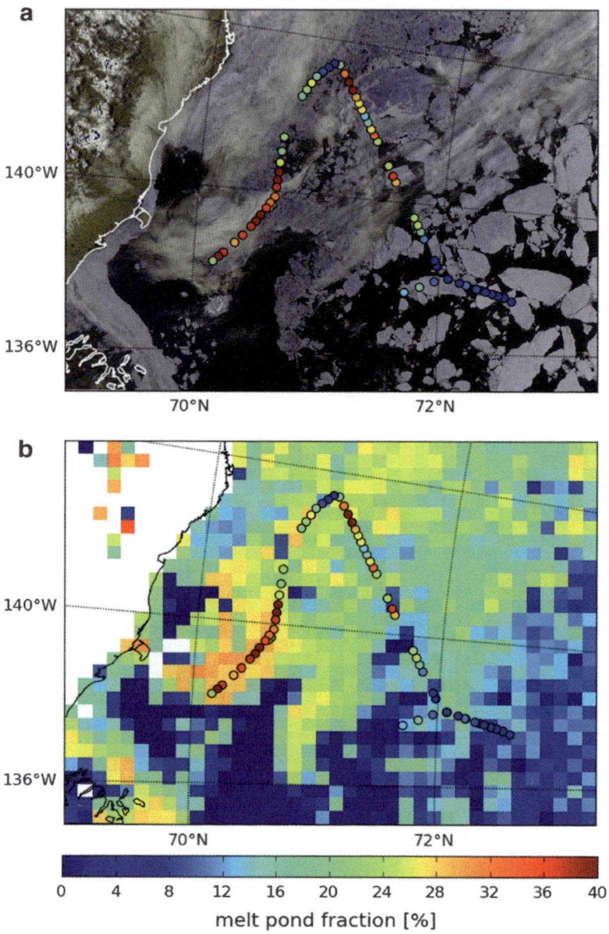

Fig. 5.6 (**a**) Spatial overlay of MELTEX melt pond results (*dots*) on a true color composite of MODIS level 1B data from June 7, 2008, 21:25 UTC. (**b**) MELTEX melt pond results (*dots*) from June 7, 2008 overlaid on MODIS melt pond fractions from mainly June 7, 2008 (detailed description in the text, Sect. 5.2.2.1)

Therefore, the spatial resolution of the melt pond set matches the resolution of the MODIS data set. For the comparison with MODIS melt pond fractions, we select the weekly MODIS data set containing the date of the high resolution acquisitions and compare the average melt pond fraction of a 12.5 km MODIS grid cell with the average of melt pond fractions of the corresponding high resolution grid cells of one 10 km × 10 km satellite scene. Locations of the three selected sites (i) Beaufort Sea, (ii) north of the Canadian Archipelago, and (iii) Fram Strait are indicated in Fig. 5.4.

Figure 5.7 shows a comparison of NSIDC melt pond fractions with MODIS melt pond data. One part of the data (mainly Beaufort data and one data point from the Canadian site) has a high agreement with MODIS melt pond fractions, the other

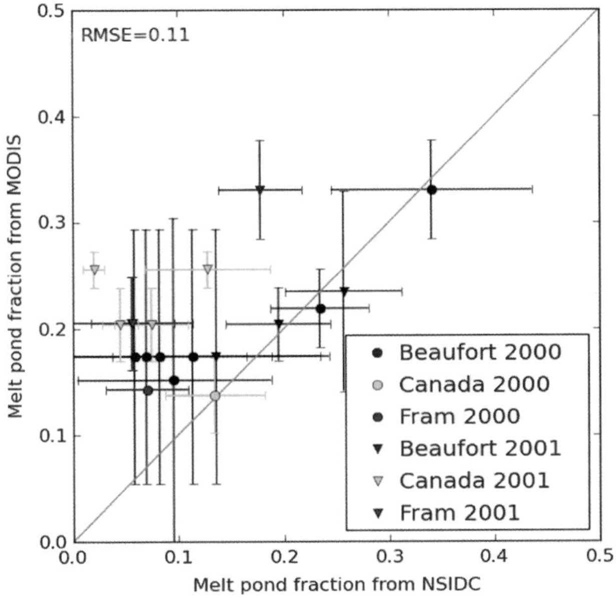

Fig. 5.7 Values from the MODIS melt pond fractions plotted versus NSIDC melt pond statistics for the years 2000 and 2001 and the three sites (i) Beaufort Sea, (ii) North of the Canadian Archipelago, (iii) and Fram Strait. The error bars are the standard deviation of the gridding from the results to 10 km × 10 km for NSIDC data and to 12.5 km × 12.5 km for the MODIS data

part of the data shows lower values than the MODIS melt pond fractions. The RMSE of the data for all sites and both years amounts 10.7 % with a coefficient of determination of $R^2 = 0.28$.

5.2.2.3 Validation with HOTRAX Data

The melt pond observations from the HOTRAX 2005 cruise [55] are also used for validation. The HOTRAX 2005 cruise track was a trans-Arctic journey, entering the pack ice on August 9, 2005 at 74°N, 160°W and traveling in the ice until its exit on September 27, 2005 at 77°N, 9°E. The expedition crossed the Bering, Chukchi, and Beaufort Seas, passed the North Pole on September 12, 2005 and exited the Arctic basin through the Fram Strait (see Fig. 5.4 on page 45).

During the cruise, observations of sea ice concentration, melt pond fraction, and ice thickness were conducted from the bridge on an hourly basis and subsequently daily mean values were calculated [55]. For this study we only use the melt pond observations. Figure 5.8 displays observation data versus MODIS melt pond fractions.

The compared mean values are distributed evenly with a coefficient of determination of $R^2 = 0.45$ and a determined RMSE of 3.8 %. The observed

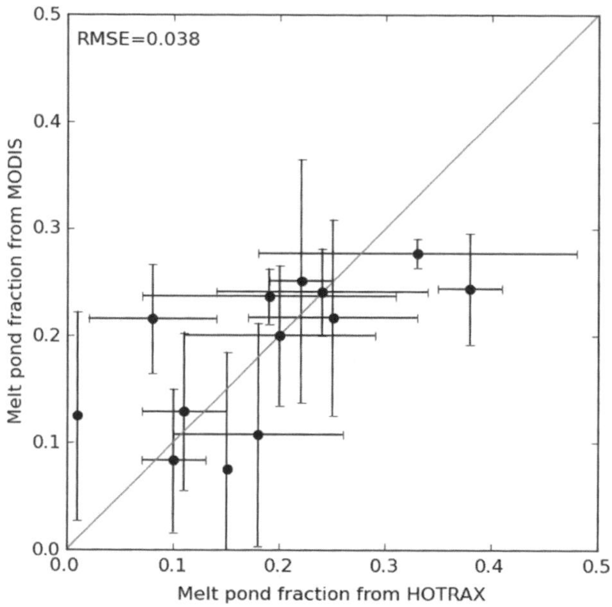

Fig. 5.8 Values of relative MODIS melt pond fractions plotted versus ship observations from the HOTRAX cruise in 2005. The error bars are the standard deviation of the daily mean for the HOTRAX data and the gridding to 12.5 km × 12.5 km for the MODIS data

differences may result from drifting sea ice, occurring due to time differences between ship observations and satellite acquisitions. There is no additional information about the accuracy of the ship observation data.

5.2.3 Resulting Melt Pond Fractions and Sea Ice Concentrations

The output layer of the ANN delivers fractions of three surface types: melt ponds, open water, as well as snow and ice for each grid cell (see Fig. 5.3 on page 43). To obtain the relative melt pond fraction on Arctic sea ice (in further context called relative melt pond fraction), we scale the melt pond data set with the sea ice concentration, created by using the MODIS open water fraction according to Eq. (5.4). To gain absolute values of the total melt pond area on Arctic sea ice (in further context called absolute melt pond area), relative melt pond fraction is scaled with the corresponding sea ice area for each year, provided by the NSIDC [20]. Note, that in this case we use the sea ice area from NSIDC and not from the MODIS sea ice concentrations, because the latter contain data gaps due to clouds and unavailable original data.

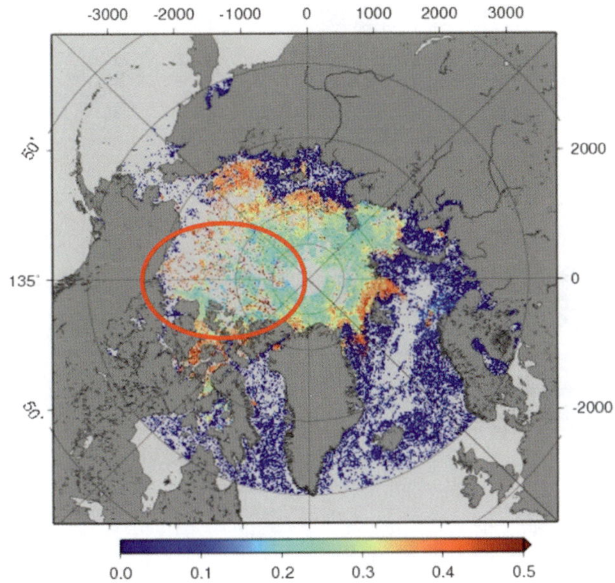

Fig. 5.9 MODIS melt pond fractions on a 500 m grid from July 23, 2003. This product is the result of the output layer of the ANN and is used as basis for the 12.5 km grid. The *red circle* indicates an area containing probably misclassified pixels

Additionally, we select only grid cells that contain more than 50 % cloud free pixels (Fig. 5.1 on page 40) to eliminate the influence of highly uncertain pixels (e.g. caused by an unreliable cloud mask, see also Sect. 5.3.1).

In the 500 m resolution pre-product, areas that contain single pixels with very high melt pond signatures are detected especially at the edges of the cloud mask (Fig. 5.9). We assume, that these pixels are a misclassified cloud signal. If we compare Fig. 5.1 on page 40 with Fig. 5.9, it is obvious, that the pixels inside the area of the red circle are as well in a region where only a few pixels involved into the coarse 12.5 km grid—if these pixels are "faulty", they can strongly influence the values of final melt pond data set.

Figure 5.10 shows the difference of a 12.5 km gridded MODIS melt pond data set of July 23, 2003 before and after masking. The high melt pond values in the Beaufort Sea result from the misclassified cloud signal (Fig. 5.9) and are eliminated after applying the pixel-weight-mask based on a 50 % threshold.

5.2.3.1 Arctic Sea Ice Concentration

Appendix A.1 on page 82 shows the MODIS sea ice concentration data set for the seasonal cycles of the years 2000–2011 in a 12.5 km grid. The years 2000, 2001, 2002, and 2007 contain data gaps. These are already existent in the initial MODIS

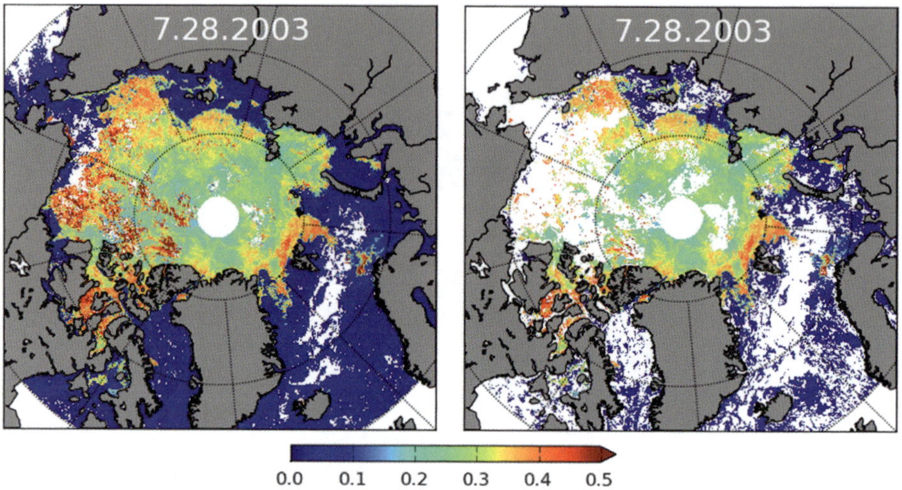

Fig. 5.10 Comparison of unmasked (*left*) and masked (*right*) data set from July 23, 2003. For masking the pixel-weight-mask based on a 50 % threshold (see Fig. 5.1) is applied

product. In 2001 one weekly data set of June 18 is completely missing. As example, Fig. 5.11 displays the weekly results of the MODIS sea ice concentration for the year 2011.

5.2.3.2 Melt Ponds on Arctic Sea Ice

Appendix A.2 on page 94 shows the MODIS melt pond data set for the seasonal cycles of the years 2000–2011 in a 12.5 km grid. The years 2000, 2001, 2002, and 2007 contain data gaps. These are already existent in the initial MODIS product. In 2001 one weekly data set of June 18 is completely missing. For further analysis, we interpolate over this missing data set. In the years 2000, 2001, 2002, and 2007, some of the weekly data sets contain no data above 80°N. After conducting test studies in advance to identify the influence of these missing data for the time series analysis, we decide to ignore the missing data, since the deviation of the relative mean melt pond fraction for the entire Arctic is below 1 % when the data gaps are substituted with the mean values of the other years.

We define a seasonal cycle from the beginning of May (day 129 of the MODIS data set) until the first week of September (day 249 of the MODIS data set).

Later in September, when new ice is formed, the used three-class model cannot resolve the spectral signature of thin ice properly and assigns this signature to the melt pond class. Therefore the algorithm specifies in autumn, when new ice is formed, areas at the sea ice edges where thin ice areas most likely occur as ponded sea ice. For this reason, no MODIS data after day 249 is analyzed.

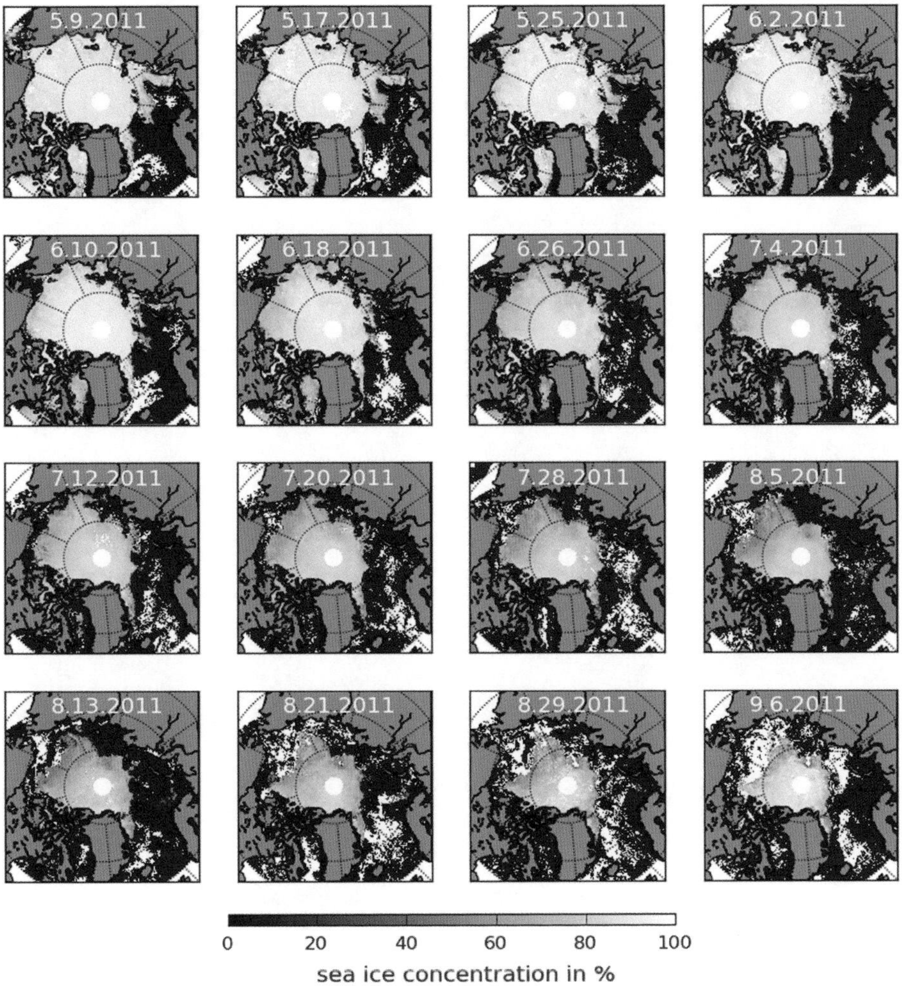

Fig. 5.11 Seasonal cycle of the sea ice concentration from MODIS satellite data in the Arctic for the year 2011

As example, the weekly results of the MODIS melt pond fraction for the year 2011 are presented in Fig. 5.12.

The here introduced melt pond data set is provided through the Integrated Climate Data Center (ICDC)[4] since January 2012.

---

[4]http://icdc.zmaw.de/cryosphere.html, accessed in February 2012.

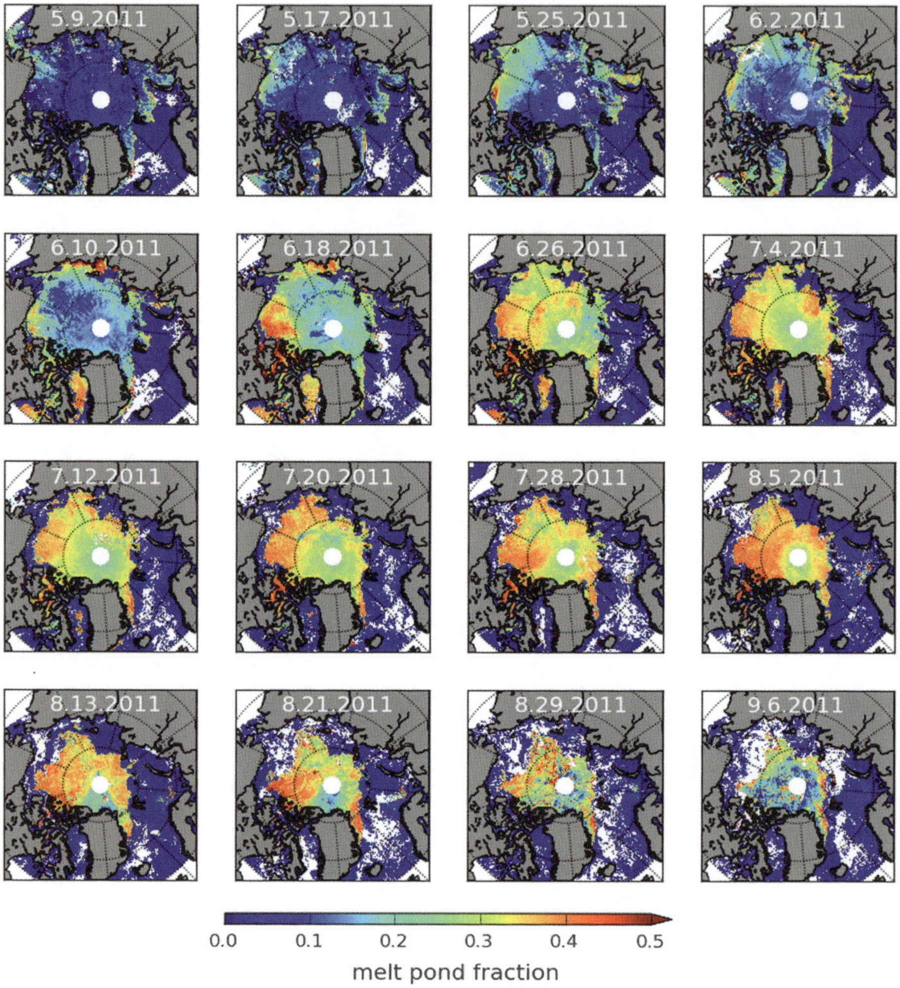

Fig. 5.12 Seasonal cycle of the melt pond fraction from MODIS satellite data in the Arctic for the year 2011

5.2.4 Melt Ponds on Arctic Sea Ice from 2000 to 2011 with Focus on the Years 2007 and 2011

During the 12 years' time period of existing melt pond data sets, two extreme minimum annual sea ice extents were recorded in the years 2007 and 2011 (see Appendix A.1 on page 82). In the following evaluation, we analyze the seasonal evolution of melt ponds, focusing on the melt pond fraction during the extreme events in 2007 and 2011.

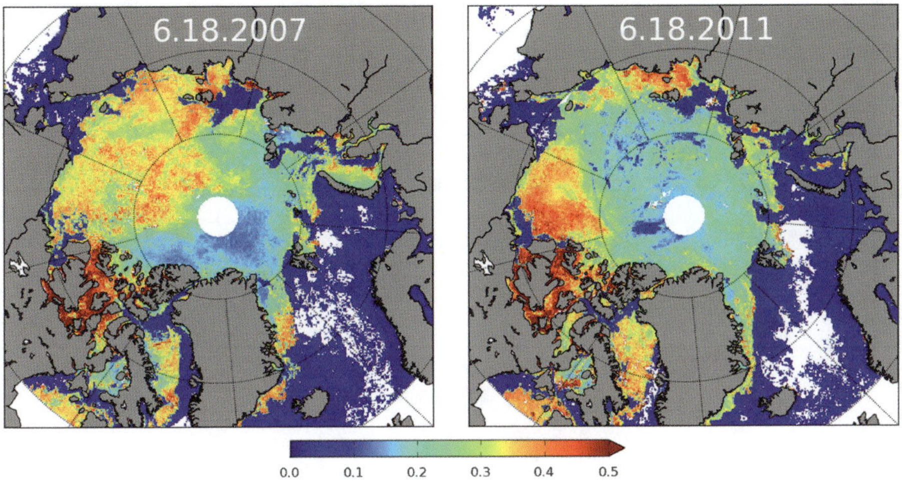

Fig. 5.13 Comparison of the relative melt pond distribution from the data sets of June 18, 2007 (*left*) and June 18, 2011 (*right*). White pixels represent missing data

In all seasonal cycles, melting features at the beginning of the melt season appear only at the sea ice edges in Greenland Sea, Kara Sea, Barents Sea, Bering Strait, Davis Strait, and Hudson Bay. The melt pond fraction continuously increases, and in mid-June an intense rise of the melt pond fraction in the Canadian Archipelago and the Beaufort Sea is observed (see also Fig. 5.13).

A comparison of the spatial melt pond distribution from June 18, 2011 with the same week in 2007 (Fig. 5.13) shows that highest melt pond fractions above 40 % occur in both years in the Canadian Archipelago. In contrast, melt pond fractions with values above 30 % appear in different regions when comparing both years: In 2007, an expansion of melt ponds mainly occurs in first-year ice areas—namely in the Beaufort Sea, Chukchi Sea, and Laptev Sea, whereas in 2011, an intensive melt pond formation is concentrated in southern parts of Beaufort Sea and Laptev Sea, as well as Baffin Bay and Hudson Bay.

Areas with melt pond fractions less than 20 % appear in 2007 only in multi-year ice regions north of the Canadian Archipelago and Greenland. In contrast, in 2011 melt pond fractions with values below 20 % are spreading from the northern Beaufort Sea over Chukchi Sea to Laptev Sea, and are also observed in the Central Arctic.

The areal extent of melt pond fractions increases in the following weeks (Fig. 5.12) and melt pond development starts at higher latitudes. However, at the same time, the total sea ice area is declining—therefore we consider both, relative melt pond fractions and absolute melt pond area (see Fig. 5.14a, b, as well as Sect. 5.2.3).

Figure 5.14a illustrates multi-annual mean relative melt pond fractions of the Arctic Ocean and the individual annual cycles for the period 2000–2011. The years

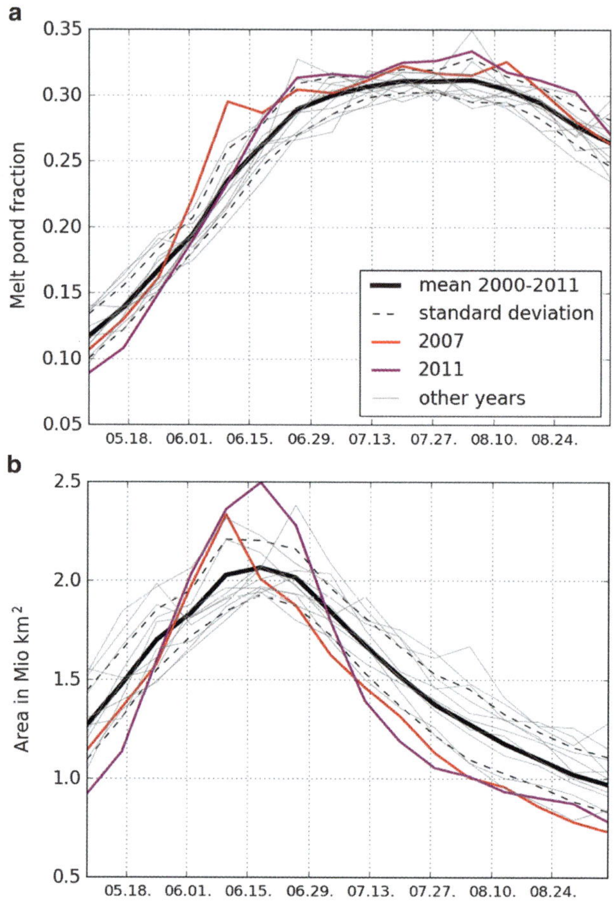

Fig. 5.14 (**a**) Mean melt pond fraction (*black line*) and standard deviation (*dashed line*) and (**b**) mean sea ice area covered with melt ponds (*black line*) and standard deviation (*dashed line*) for the time period 2000–2011 for the entire Arctic. The light gray lines display the development of melt ponds for all years in the specified time period. The years 2007 (*red*) and 2011 (*magenta*) are displayed separately for comparison

2007 and 2011 are highlighted in red and magenta, respectively. The relative mean melt pond fraction shows a strong increase in June with a maximum of 30 % lasting from end of June until beginning of August. Seasonal cycles of the years 2007 and 2011 begin both below the average curve of the relative melt pond fraction. In 2007, the relative melt pond fraction has its highest increase and rises above average values in the beginning of June. A first maximum of relative melt pond fraction in 2007 is reached in mid-June, which is followed by above-average values in July and August. In contrast to 2007, the relative melt pond fraction shows in 2011 a smoother increase at the beginning of the melt season and exceeds the mean values in mid-June. From this point in time the curve of 2011 remains up to two standard deviations above average (30 ± 1.2 %) and even above the high values of 2007.

Figure 5.14b depicts absolute melt pond areas of the Arctic region for all years. It is evident that the annual sea ice extent has a strong influence on total melt pond area. Both years, 2007 and 2011 are characterized by a below average melt pond area until the end of May, which can be attributed to negative sea ice concentration anomalies. Additionally, a strong increase of melt ponds in June and their early occurrence have a severe influence on the extent of the total melt pond area. In 2011, the observed maximum melt pond area is 2.5 million km^2, which is 0.4 million km^2 more than the maximum of all years. With 2.4 million km^2, the 2007 maximum is slightly lower than 2011, but is still more than one standard deviation above the average value of 2.1 million km^2. Note that for both years a similar melt pond growth rate is found until the 2007 maximum is reached in the first week of June. The total melt pond area for 2011 increases further until mid-June. Due to the outstanding decrease of the sea ice area in 2007 and 2011, the total melt pond areas drop below the average value in both years in end of June.

Figure 5.15a displays a Hovmoeller diagram of the zonal mean of melt ponds for the last 12 years. It illustrates a dependence of the temporal development of melt ponds from the geographical latitude. The maximum of the averaged relative melt pond fraction occurring from mid-June to mid-August is located at the latitudes between 70° and 80°N. A second maximum emerging in June at lower latitudes (60°–62°N) indicates the coastal melting and the melt ponds in Hudson and Baffin Bay. The patterns of the anomalies for 2007 and 2011 look very similar (compare Figs. 5.15c and d), where both exhibit higher relative melt pond fractions at high latitudes than the mean reference period (Fig. 5.15a), and in 2011 melting at low latitudes is more pronounced.

As can be seen from Fig. 5.14a, in 2007 largest differences from the mean values occur at the beginning of the melt season. In the years 2007 and 2011, strong positive anomalies of relative melt pond fractions occur between 80° and 90°N, especially in mid-June (see Fig. 5.15c, d). Positive significant trends of up to 3 % at a 95 % significance-level can be identified in June and August in Fig. 5.15b. For the trend calculations we define the melt pond fraction values of the year 2000 as 100 % and calculate the deviation from this value for the following 11 years. These positive trends reflect the increase in the relative melt pond fraction. The negative trends of −4 % in August below 70°N depict the declining sea ice extent.

Figure 5.16 on page 59 shows the time series of the mean relative melt pond fraction and the total melt pond area for the entire time range from 2000 to 2011. The mean relative melt pond fraction of all years is 25.1 ± 1.7 %, the mean total melt pond area is 1.49 ± 0.11 million km^2 and shows a negative trend for the entire time range. Although the time series composed of only twelve values is very short, the calculated trend of −16.4 % for the decreasing mean melt pond area is statistically significant at the 90 % significance-level, proved with the Cox and Stuart test. For 2011, the mean melt pond area with 1.42 million km^2 is very close to the mean value, although the overall maximum of the melt pond area was recorded in mid-June that year. However, the low sea ice extent present in late summer, compensates

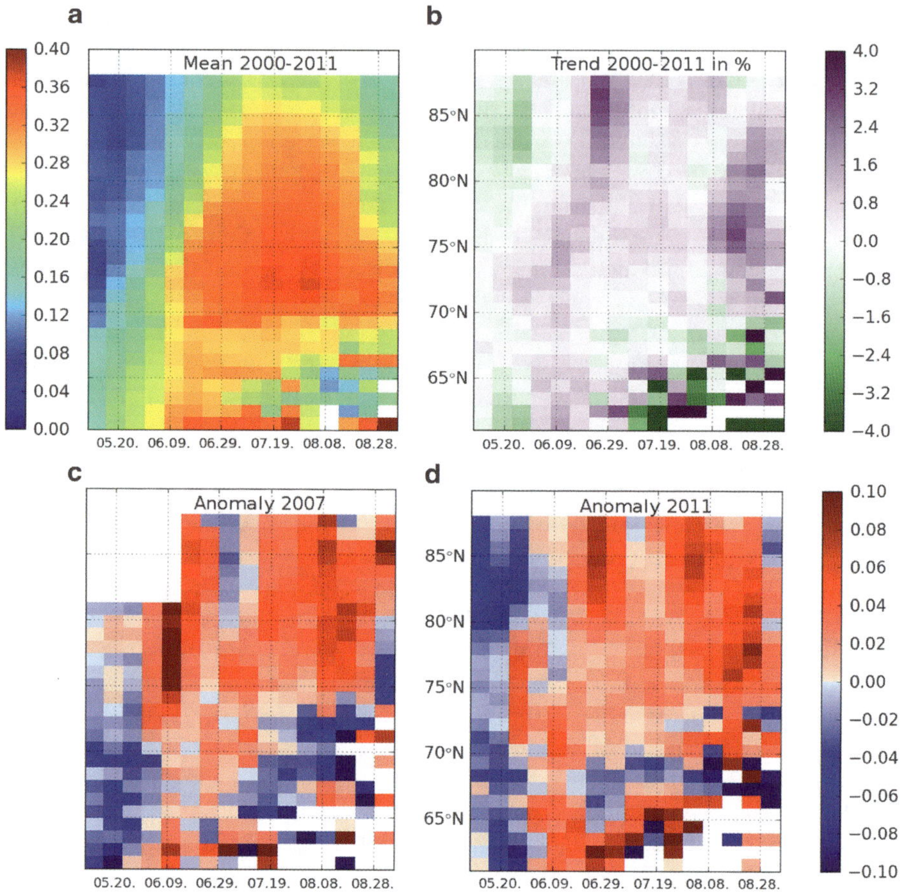

Fig. 5.15 (**a**) Zonal mean of relative melt pond fraction over the entire Arctic from 2000 to 2011; (**b**) trends of zonal relative melt pond fraction over the period from 2000 to 2011; (**c**) anomaly of relative melt pond fraction in 2007; (**d**) anomaly of relative melt pond fraction in 2011

the extreme values in early summer. The relative melt pond fraction exhibits a non-significant, slight upward trend of 2.4 %, where the years 2007, 2010 and 2011 record the three highest values of the time series.

5.2.5 *Influence of Melt Ponds on Microwave Sea Ice Concentration Retrievals*

Sea ice concentration retrievals on passive microwave sensors like SSM/I (Special Sensor Microwave Imager) or AMSR-E (Advanced Microwave Scanning Radiometer for EOS) have been used since more than three decades and are

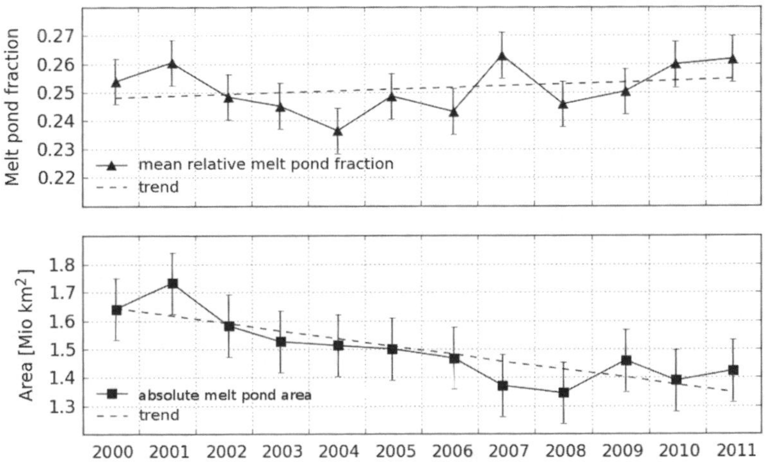

Fig. 5.16 Mean relative melt pond fraction (*top*) and mean absolute melt pond area (*bottom*) for the years 2000–2011 for the entire Arctic with its standard deviations and trends

well-investigated and reliable on a large scale. Nevertheless, the influence of melting sea ice, especially of melt ponds on the sea ice surface, causes uncertainties in all retrieval algorithms [7, 12, 74].

With MODIS melt pond fractions, we have now the ability to compare melting features directly with sea ice concentrations. Additionally, the obtained MODIS sea ice concentrations are useful to estimate uncertainties in microwave retrievals.

In this context, a case study in the Canadian Archipelago on the data set of June 25, 2008 is performed. The study area is a 250 km × 100 km region around the coordinates 72°N and 110°W (see Fig. 5.17 on page 60). Very high relative melt pond fractions with values up to 70 % occur in June on the flat level ice in this region (see Fig. 5.17 bottom).

The first three images of Fig. 5.17 display sea ice concentrations from the AMSR-E sensor, calculated with the ASI algorithm [73], the NASA-Team 2 (NT2) algorithm [42], and the Bootstrap (BT) algorithm [10] for the area of the Canadian Archipelago. Sea ice concentrations range in all three cases from 50 % to 80 %. The fourth image shows MODIS sea ice concentrations. They indicate values from 90 % to 100 %.

The ASI algorithm yields a mean sea ice concentration of 45 %; the Bootstrap algorithm yields for the same area 55 %, and the NASA Team 2 algorithm shows a value of 56 %. Compared to the sea ice concentration of 93 % retrieved from the MODIS data, all microwave algorithms underestimate the sea ice concentration by around 40 %. It is evident, that the areas containing high melt pond fractions (>50 %) exhibit low sea ice concentrations for the microwave retrieved results. The MODIS sea ice concentrations show an homogeneous distribution around 90 %.

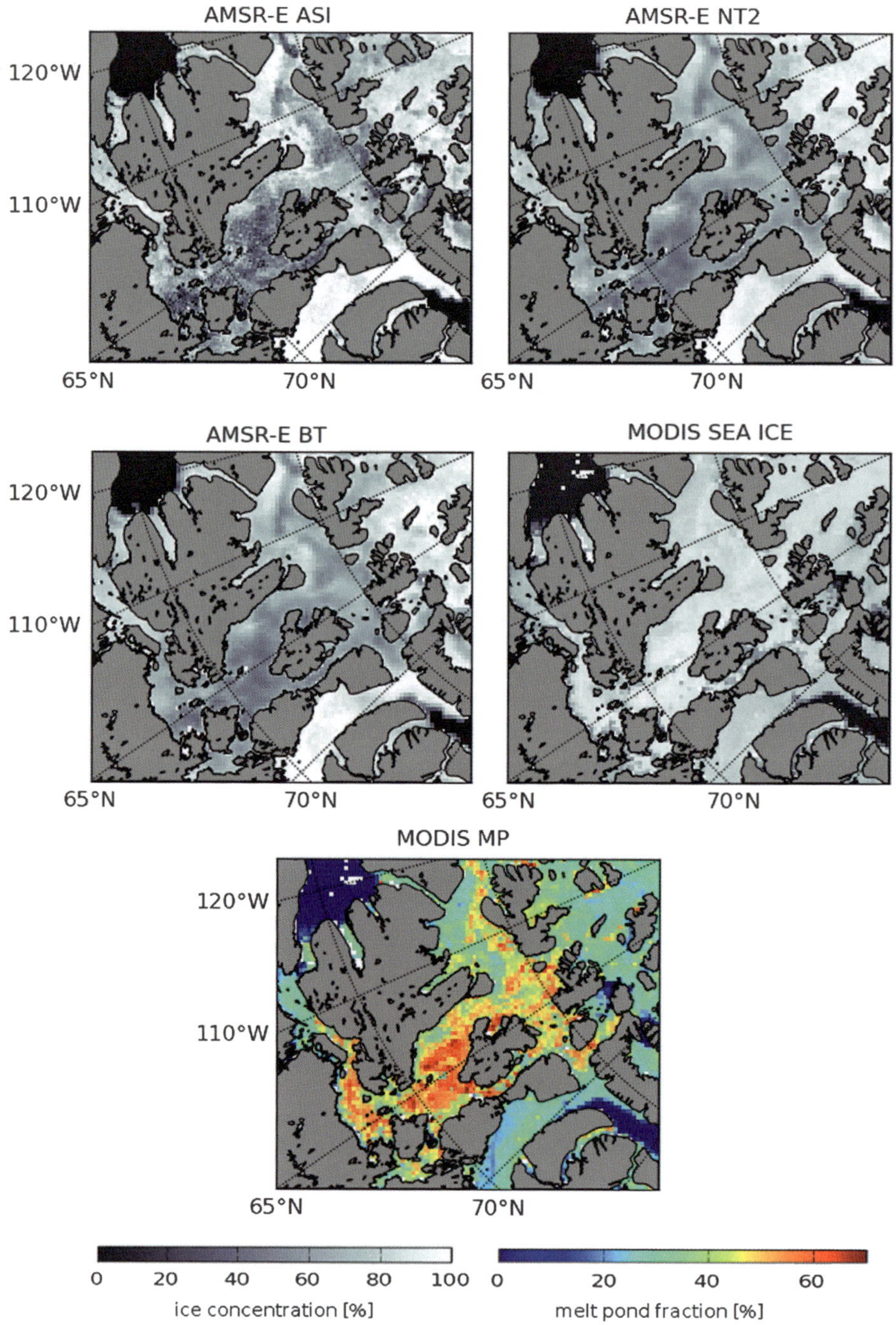

Fig. 5.17 Comparison of AMSR-E ASI (*top left*), AMSR-E NASA Team 2 (*top right*), AMSR-E Bootstrap (*middle left*), and MODIS (*middle right*) sea ice concentrations. *Bottom figure* shows the MODIS melt pond fraction. All images are from May 25, 2008

5.3 Discussion and Conclusion

5.3.1 Discussion

The determination of the melt pond fraction on a large scale over a multi-annual time period is important to estimate the contribution of melt ponds on the ice-albedo feedback mechanism. The method described in Sect. 5.1 is suitable to derive daily and weekly data sets of the melt pond fraction from the MODIS surface reflectance product. Additionally, sea ice concentrations from the same MODIS product are obtained. Daily data sets are used to validate MODIS melt pond fractions, and from weekly data sets we generate time series for the entire Arctic.

The weekly data set is a composition of selected pixels from daily acquisitions, especially those with a minor cloud cover fraction and other atmospheric influences. Therefore, this product is a patchwork of conditions for a specific day within one week.

The comparison of daily MODIS melt pond fractions with melt pond fractions derived from aerial photos and sea ice observations from different locations in the Arctic indicates a high accordance. This confirms that the technique to estimate the melt pond fraction performs well.

The validation with the NSIDC melt pond data set results in generally higher MODIS melt pond values and a poor coefficient of determination with $R^2 = 0.28$. In this case we compare the average melt pond fraction of a 10 km × 10 km scene from NSIDC with a 12.5 km × 12.5 km grid cell of the MODIS melt pond fraction. Here, we use weekly data sets for validation instead of daily data sets. We have noticed that in the daily data sets the validation area is poorly covered by valid pixels, whereas the weekly data sets provide more cloud free pixels. However, this leads to the problem, that the according pixels in the MODIS data are probably from a different date than the NSIDC data. This might be a reason for the higher values of the MODIS melt pond fractions.

The melt pond fractions identified from aerial photos of the MELTEX data set with a high spatial resolution have a higher variability than those from the MODIS melt pond data set. This may result from the relatively coarse resolution of 500 m per MODIS pixel, used as the basis for this product, while the average size of melt ponds is about 15–60 m^2 [61].

Potential sources of errors are to be assumed in the atmospheric correction routines and in the influences of the viewing angles as well as the solar geometry. The BRDF correction of the MOD09 product is not applied on all areas of the Arctic sea ice, especially not over the deep ocean, because due to moving and rapidly changing sea ice surface the BRDF determination for a specific area is difficult. For most of the areas, model results for first-year and multi-year ice are used as "a priori" estimates of the BRDF.

The assumption of a three-surface-class model also causes uncertainties regarding the different surface fractions, since the Arctic sea ice cover is actually a mosaic of various surface types—and not only melt ponds, snow and ice, and open water. The method described here is based on satellite data from optical sensors. Therefore, melt pond fractions can only be identified from cloud free data. The cloud mask, integrated into the MOD09 product, does not capture all actual present clouds. Especially over highly reflecting surfaces like sea ice, the used cloud mask may have problems. It may occur, that low-lying clouds cannot be identified and appear as highly reflecting surface features.[5] To improve data quality for further studies, a data mask with a defined threshold or weighting of involved pixels can be applied after the gridding routine. In the 500 m grid, single pixel with a high melt pond signal occur often within cloudy fields or at the edges of the cloud mask. We assume that these pixels are a misclassified cloud signal. To minimize the influence of these misclassified pixel, the above introduced technique eliminates at least grid cells with a low data quality and assures that each 12.5 km grid cells contains at least 50 % cloudfree pixels. However, the existing problem of cloudy pixels in the initial dataset can impact the melt pond fraction and should be considered. Nevertheless, sufficient reasons remain to investigate more in a useful cloud detection.

In the autumn results, the used three-class model with open water, melt pond, and snow and ice fractions cannot resolve the spectral signature of thin ice properly and assigns this signature to the melt pond class. Therefore, the algorithm specifies in autumn, when new ice is formed, areas at the sea ice edges where thin ice areas most likely occur as ponded sea ice. For this reason the melt pond data set ranges from the beginning of May (day 129) only until the first week of September (day 249).

From the results as displayed in Fig. 5.13 on page 55, it is apparent that the spatial pattern of the melt pond distribution correlates with atmospheric conditions: In 2007, a persistent dipole anomaly weather pattern with unusually high pressure over the Beaufort Sea and a low pressure system over central and western Siberia [75] caused strong southerly winds from the Bering Straight across the North Pole, which transported warm air and water masses into the Arctic. Additionally, a high pressure system over the Beaufort and Chukchi Sea, formed in early June 2007 and persisted for 3 months, which caused a predominately clear sky [75], has been allowed more incoming shortwave radiation. The negative cloud cover anomaly in the Beaufort Sea region for June–August 2007 exhibits values of up to −25 % [68]. Figure 5.13 on page 55 displays an extensive melt during June to August in the same regions. In 2011, a similar pressure field, but less strong and persistent as in 2007, was observed. The location of the high pressure center is shifted towards the north of Greenland and the low pressure center is shifted towards Alaska, causing easterly winds rather than southerly winds as observed in 2007.[6] The distribution of melt ponds in Fig. 5.13b reflects this shift: A distinct formation of high melt pond fractions is identified in the Beaufort Sea. The cloud cover anomalies in July 2011 as

[5] Interactive comment from Marc Tschudi on The Cryosphere Discussion, 5, 2991, 2011.

[6] NCEP Reanalysis data from http://www.esrl.noaa.gov/psd, accessed in February 2012.

presented by Overland (2011) [53], [7] display values of up to −20 % and are located over the Beaufort Sea and the Central Arctic.

Figure 5.15 on page 58 shows a zonal increase of the melt pond fraction correlated to the increasing solar elevation. The maximum of the average melt pond fraction is located at the higher latitudes, between 70° and 80°N, although the melt period is shorter at high latitudes. This was also observed by Tschudi et al. (2008) [81] and may be caused by a larger sea ice concentration and a thicker ice coverage, which allows less runoff. Therefore, more water is accumulated to form melt ponds [81].

The pattern of melt pond distribution in early summer, characterized by areas with an increased melt pond fraction in the beginning of June, indicates the ice free areas later in September (see Fig. 6.2 on page 94). This is in good agreement with the statement of Perovich (2002, 2011) [59, 61], that an early occurrence of melt ponds has a strong influence on the formation of open water areas. Clearly visible is also the formation of homogeneous areas with a very high melt pond fraction up to 70 % at the end of June on the flat first-year ice in the Canadian Archipelago. The decrease of the melt pond fraction is starting in most of the years in end of August.

The downward trend of the Arctic sea ice extent and the two sea ice minimum records of the years 2007 and 2011 provoke a debate about the possible causes of the recent decline of the Arctic sea ice area. A thinner ice cover is described as more susceptible to the enhanced summer melt [40]. Additionally, the loss of multi-year sea ice and the associated increase of the first-year ice fraction [47] enhance the melt pond formation [2, 15].

It is obvious that the low sea ice extents recorded in the last years are caused by a combination of a high melt pond fraction at an early stage of the melting cycle and a thinning Arctic sea ice coverage which is mainly composed of first-year ice.

5.3.2 Conclusion

Here, we present a novel method to retrieve a multi-annual data set of melt pond fractions and sea ice concentrations derived from MODIS satellite data for the entire Arctic. Both data sets comprise the last twelve melt cycles and thus provide a profound basis for further studies.

The MODIS surface reflectance values are described by an overdetermined set of linear equations, which are not well conditioned. These equations are solved with an optimization method. As result, three surface fractions are obtained: open water, snow and ice, and melt ponds. To constrain the interval of the solution between zero and one, a cost function is implemented.

[7] http://www.arctic.noaa.gov/repordcard/temperature_clouds.html, accessed in February 2012.

To accelerate the processing, a trained ANN is used. Performance tests of the ANN as described in Sect. 5.2.1 yield stable results with a maximal error of +6 % for melt pond fractions.

The validation of relative melt pond fractions is conducted with three different data types from local observations: Aerial photos from the MELTEX campaign in the Beaufort Sea in June 2008 are compared with MODIS melt pond fractions. The RMSE of the comparison are 11.2 % and 10.6 %, respectively. Additionally, we compare analyzed intelligence satellite data with a resolution of 1 m × 1 m of three sites in the Arctic Ocean for the years 2000 and 2001. The RMSE for the data of all sites and both years amounts 10.7 %.

Finally, the validation with ship observation data from the HOTRAX 2005 cruise through the Arctic Ocean results in an RMSE of 3.8 %.

The mean relative melt pond fraction for the entire Arctic derived from MODIS satellite data of the last 12 years shows a strong increase in June with a broad maximum of 30 % appearing in July and August. In contrast, the averaged absolute melt pond area has its maximum of 2.1 million km^2 in mid-June.

The analysis of annual relative melt pond fractions yields a negative trend of −16.4 % of the total melt pond area over the entire melt season, corresponding to a declining sea ice extent. In the years of extreme sea ice loss, namely 2007 and 2011, maxima of total melt pond area that are situated one, respectively two standard deviations above the average value of 2.1 million km^2 can be observed. From the temporally and spatially resolved trends (Fig. 5.15d), an increase of the melt pond fraction from 80°–88°N in June and August is identifiable, with a maximum at the beginning of the melt season in June and a second maximum at the end of the melt season in August. This in turn leads to a prolongation of the melt season and is in agreement with the studies of Markus et al. (2009) [45].

Additionally, we show in a case study that the influence of surface melting on sea ice concentrations as retrieved from passive microwave data causes an underestimation of values of up to 40 %, compared to the MODIS sea ice concentration results.

This study shows the importance to further analyze the early appearance of melt ponds that potentially causes enhanced melting.

Chapter 6
Summary and Outlook

6.1 Summary

The strong decline of the Arctic sea ice extent, especially in the last decade, and the inability of recent climate models to predict this development, give a strong motivation to collect observational data to better understand the interaction between atmosphere, sea ice, and ocean. Because melt ponds have a significant influence on the amount of sea ice melt and therefore also on Earth's radiation balance, the primary goal of this study was to develop a multi-annual melt pond data set of the entire Arctic region based on satellite data. On the basis of such a melt pond data set it is possible to answer the following research questions:

- What is the mean annual melt pond fraction of the Arctic and how does a seasonal cycle look like?
- Is there a relation of melt pond fraction and sea ice extent minimum?
- Do melt ponds influence sea ice concentration values as retrieved from passive microwave sensors?

Markus et al. (2003) [44] have presented the potential of Landsat data to determine melt ponds on Arctic sea ice in a case study [44]. As the USGS Landsat archive became freely accessible in 2009, we applied the determination method that is proposed by Markus et al. (2003) [44] and found, that this method has limitations regarding a potential automatic processing routine: Tie-points and thresholds for the identification of melt ponds need to be adjusted for each scene separately. We have developed a new method, using principal component analysis (PCA) for the determination of melt ponds and compared the achieved results with Markus' method. The coefficient of determination is $R^2 = 0.94$, the melt pond fractions of both methods differ by about 15 % [63].

A subsequent archive search yielded, that only in individual cases Landsat satellite images can be used for our purposes, since they are very sensitive to highly reflecting surfaces like sea ice. Nearly all scenes that contain melt ponds

A. Rösel, *Detection of Melt Ponds on Arctic Sea Ice with Optical Satellite Data*, Hamburg Studies on Maritime Affairs 25, DOI 10.1007/978-3-642-37033-5_6,

have problems with saturated pixels. Saturation is related to sun elevation and the surface type. Unfortunately, the melt pond fraction increases with an increasing sun elevation angle. Additionally, the data coverage is poor above the ocean. Therefore, satellite imagery from Landsat sensors is inapplicable as a data source for large scale applications.

However, the methods introduced here for the melt pond determination from high resolution spectral sensors like Landsat can be used to obtain results from case studies or to compare the results with the findings of other studies.

The main goal was successfully achieved by using medium resolution spectral satellite data from the MODIS sensor. Motivated by the approach of Tschudi et al. (2008) [81], we set up a system of linear equations, which bases on the MODIS surface reflectance values. The equations describe the composition of three surface types on the sea ice, namely open water, melt pond, and snow and ice, based on their spectral differences. The overdetermined system of linear equations was solved with an approximation method. To comply with the physical principles, it was necessary to constrain the solution to an interval between zero and one with a cost function as side condition. To evade the high computational costs that were caused by this solution statement and to speed up processing, an artificial neural network (ANN) was successfully trained. With this approach, we have obtained a multi-annual melt pond data set of the entire Arctic from MODIS data for the first time [65].

We have validated the relative melt pond fractions with three different data types from local observations: firstly, we have used aerial photos from the MELTEX campaign in the Beaufort Sea in June 2008. The RMSE of the comparison of two MELTEX data sets with two MODIS melt pond sets were 11.2 % and 10.6 %, respectively. Secondly, we have compared analyzed intelligence satellite data with a resolution of $1\,\mathrm{m} \times 1\,\mathrm{m}$ of three sites in the Arctic Ocean for the years 2000 and 2001. The RMSE for the data of all sites and both years amounts to 10.7 %.

Thirdly, ship observation data from the HOTRAX 2005 cruise in the Arctic Ocean were used for validation. The RMSE between ship and satellite data resulted in 3.8 %.

A time series analysis of the MODIS melt pond data set resulted in average values for the relative melt pond fraction of $25.1 \pm 1.7\,\%$ and an average area of 1.49 ± 0.11 million km^2 for the entire Arctic for the time period from 2000 to 2011. The mean relative melt pond fraction begins to raise in mid-may, has a strong increase in June, and shows a broad maximum around 30 % in July until the beginning of August. Thereafter, the mean relative melt pond fraction decreases. The absolute melt pond area has its average maximum of 2.1 million km^2 in mid-June. The concurrent decline of the sea ice extent causes a constant decrease of the total melt pond area until refreezing occurs in September. Additionally it was noted, that the annual total melt pond area exhibits a negative trend over the entire melt season, in accordance to the declining sea ice extent.

In the years of extreme sea ice loss, namely 2007 and 2011, we observed a maximum total melt pond area in mid-June to end June. From the temporally and spatially resolved trends (Fig. 5.15d), an increase of relative melt pond fraction from

80°–88°N in June and August is evident. This in turn leads to a prolonged melt season and agrees with the results of Markus et al. (2003) [45]. The outcome of our study provides strong evidence for the importance of the early appearance of melt ponds.

To show the influence of melt ponds on microwave based sea ice concentration retrieval algorithms, we have performed a case study, where we compared sea ice concentrations from AMSR-E sensor, calculated with the ASI algorithm [73], the NASA-Team 2 algorithm [42], and the Bootstrap algorithm [10] with MODIS sea ice concentrations. The test area in the Canadian Archipelago showed a high fraction of melt ponds (up to 70 %). Compared to the sea ice concentration of 93 % retrieved from the MODIS data, all microwave algorithms underestimate the MODIS sea ice concentration in this case by around 40 % [65].

The MODIS melt pond data set introduced here is provided through the Integrated Climate Data Center (ICDC).[1]

6.2 Outlook

The results of this study can be used to highlight aspects of sea ice surface albedo and transmittance of solar radiation through melt ponds. With the multi-annual melt pond data set, the modeling community will be able to improve the description of melt ponds in climate and ocean models for a better representation of the early summer sea ice melting and consequently the determination of the sea ice extent. The length of melt seasons as well as the inter-annual variability of melting features can be observed and trends of melt processes and their influence on the sea ice-albedo feedback mechanism can be determined. Further analysis of time series, also on a more regional scale, as well as a combination of melt pond fractions with ice drift products are prospective applications of the melt pond data set. This could lead to predictions of early ice free sea routes through the Arctic Ocean, which is of high economic interest. Additionally, a continuation of the time series in the future is essential to recognize potential trends in the development of the melt pond fraction.

As shown in Fig. 5.17 on page 60, the MODIS sea ice concentrations can be used to estimate the influence of melt ponds on the sea ice concentration determination from microwave sensors like AMSR-E. In this example, all AMSR-E algorithms clearly underestimate MODIS sea ice concentrations by around 40 %. Further studies to identify a relation between melt pond fraction and microwave-retrieved sea ice concentrations are necessary. Additionally to the melt pond data set, we achieved a sea ice concentration data set from MODIS data. This sea ice

[1] http://icdc.zmaw.de/, accessed in February 2012.

concentration data set, in combination with the melt pond data set, is a valuable basis for quantifying the error caused by melt features in sea ice concentration data derived from passive microwave imagery. However, the MODIS sea ice concentrations are not validated until now. This data set additionally carries the capacity for comparison with different microwave-retrieved sea ice concentrations and their evaluation.

Another point for future activities should be a further validation of the MODIS melt pond data set and the determination of clouds and their influence. For validation purposes, very high resolution satellite data like from Quickbird or GeoEye-1 can be used. Also the comparison with different melt pond detection techniques like from synthetic-aperture radar (SAR) and/or scatterometer data [31, 87], is conceivable. For an improvement of the cloud detection over Arctic sea ice, cloud information from different sensors like from CloudSat-CALIPSO [41], could be applied to reassess the quality of the MODIS cloud mask.

A more precise analysis regarding the number of surface classes would be worthwhile. A classification into light and dark melt ponds could result in a depth or volume estimation and hence the potential of energy transmission could be specified. Furthermore, the introduction of a bare ice and/or thin ice class would be helpful to reduce the uncertainties of the melt pond data set. The introduction of one or more additional surface classes requires a more differentiated spectral resolution of the optical sensor. For this, it is imaginable to modify this melt pond detection method for hyperspectral scanners or sensor like AVIRIS (Airborne Visible InfraRed Imaging Spectrometer) or EnMAP (Environmental Mapping and Analysis Program).[2]

Additionally, a higher spatial resolution comparable to the planned Sentinel-2 mission[3] could help to improve the results of the melt pond data set, especially in regions where a high spatial variability is present (e.g. drifting ice floes in the Fram Strait).

Furthermore, the adaption of the melt pond determination method for the sea ice coverage around Antarctica as well as for the Antarctic and Greenland ice sheets could provide further comprehension of melting processes. This could give valuable scientific insights into the role of melting processes in interaction with ice, ocean, and atmosphere and on which scale they influence the tipping elements like the Greenland or West Antarctic ice sheet as well as the Antarctic bottom water formation [14, 39].

---

[2]The EnMAP mission is managed by the Space Agency of the German Aerospace Center (DLR). The envisaged launch of the EnMAP satellite is 2015.

[3]Launch of the satellite is planned for 2013, the spatial resolution rages from 10 to 60 m.

Appendix

A.1 Satellite Specifications (Tables 6.1 and 6.2)

Table 6.1 Landsat 7 ETM+ spectral band characteristics [38]

Band	Description	Bandwidth (μm)	Resolution (m)
1	Blue	0.45–0.515	30
2	Green	0.525–0.605	30
3	Red	0.63–0.69	30
4	Near infrared	0.75–0.9	30
5	Mid infrared	1.55–1.75	30
6	Thermal	10.4–12.5	60
7	Mid infrared	2.09–2.35	30
8	Panchromatic	0.52–0.9	15

Table 6.2 Description of 36 spectral MODIS bands [48]

Primary use	Band	Bandwidth (nm)	Resolution (m)
Land/cloud/aerosols	1	459–479	500
	2	545–565	500
Land/cloud/aerosols	3	620–670	250
	2	841–876	250
Land/cloud/aerosols	5	1230–1250	500
	6	1628–1652	500
	7	2105–2155	500
Ocean color/phytoplankton/biogeochemistry	8	405–420	1,000
	9	438–448	1,000
	10	483–493	1,000
	11	526–536	1,000
	12	546–556	1,000
	13	662–672	1,000
	14	673–683	1,000
	15	743–753	1,000
	16	862–877	1,000
Atmospheric water vapor	17	890–920	1,000
	18	931–941	1,000
	19	915–965	1,000
Surface/cloud temperature	20	3.660–3.840	1,000
	21	3.929–3.989	1,000
	22	3.929–3.989	1,000
	23	4.020–4.080	1,000

(continued)

Table 6.2 (continued)

Primary use	Band	Bandwidth (nm)	Resolution (m)
Atmospheric temperature	24	4.433–4.498	1,000
	25	4.482–4.549	1,000
Cirrus clouds water vapor	26	1.360–1.390	1,000
	27	6.535–6.895	1,000
	28	7.175–7.475	1,000
Cloud properties	29	8.400–8.700	1,000
Ozone	30	9.580–9.880	1,000
Surface/cloud temperature	31	10.780–11.280	1,000
	32	11.770–12.270	1,000
Cloud top altitude	33	13.185–13.485	1,000
	34	13.485–13.785	1,000
	35	13.785–14.085	1,000
	36	14.085–14.385	1,000

A.2 MODIS Sea Ice Concentrations (Fig. 6.1)

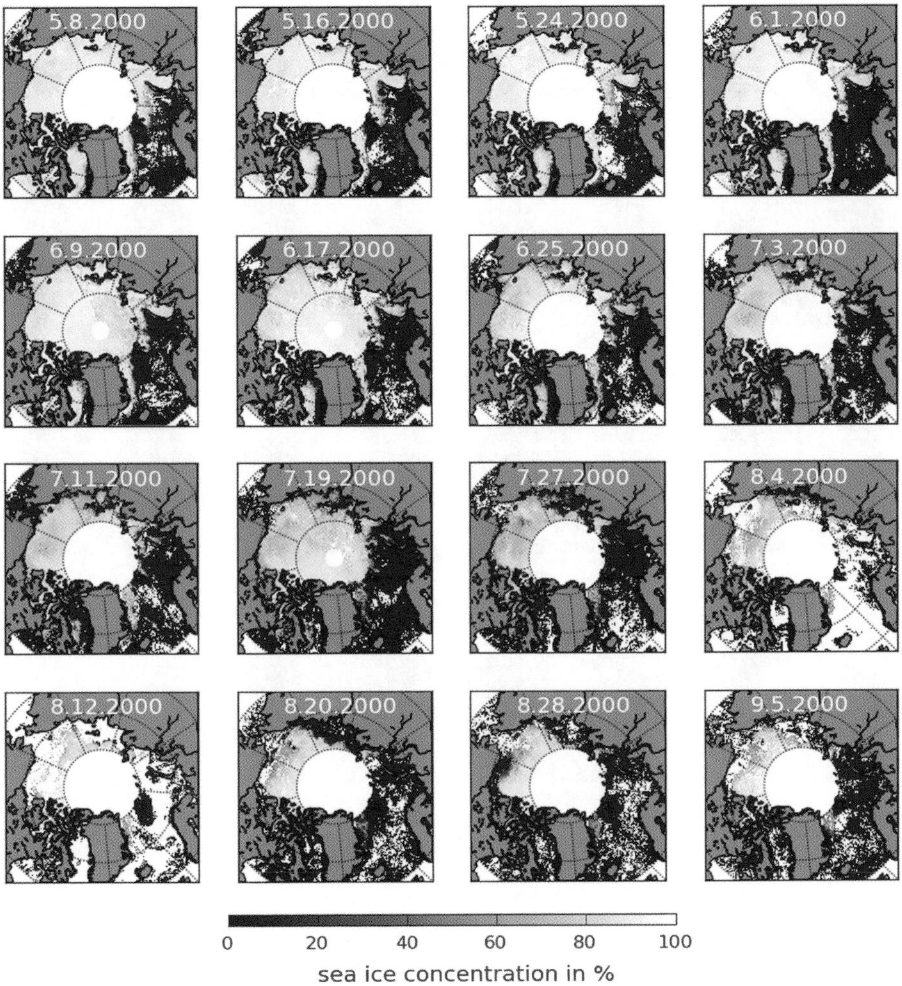

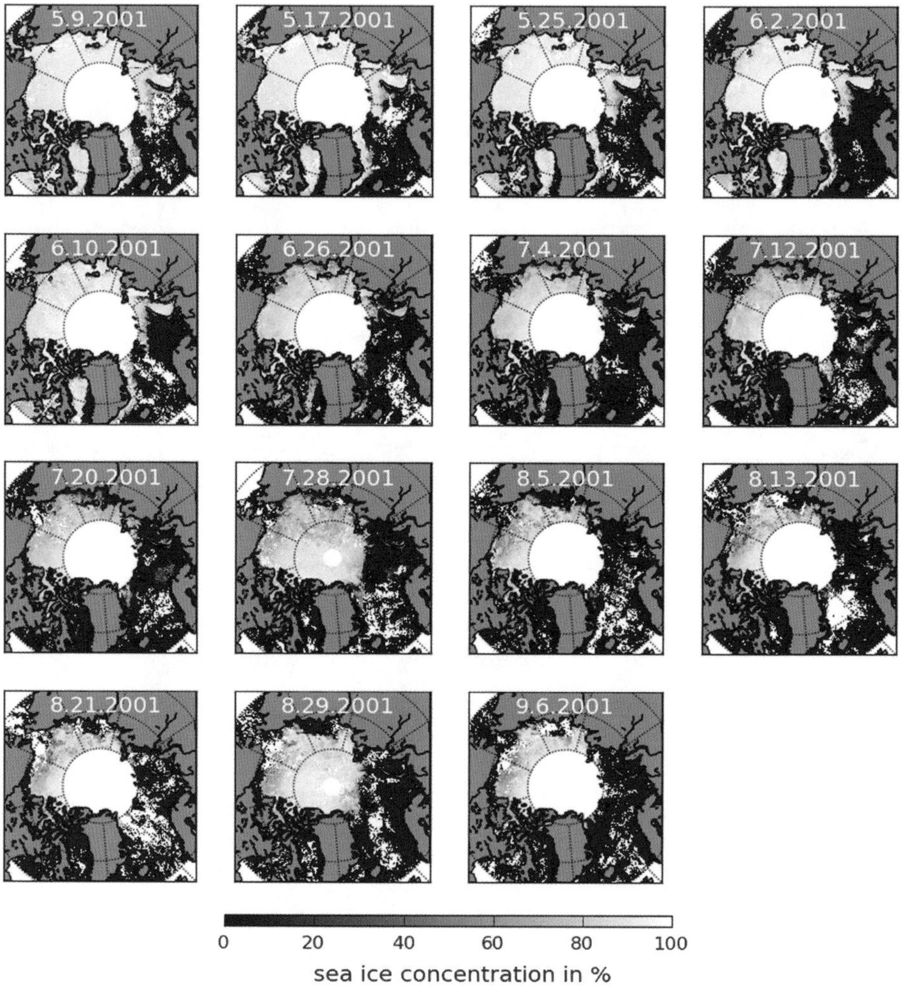
5.9.2001
5.17.2001
5.25.2001
6.2.2001
6.10.2001
6.26.2001
7.4.2001
7.12.2001
7.20.2001
7.28.2001
8.5.2001
8.13.2001
8.21.2001
8.29.2001
9.6.2001
0
20
40
60
80
100
sea ice concentration in %

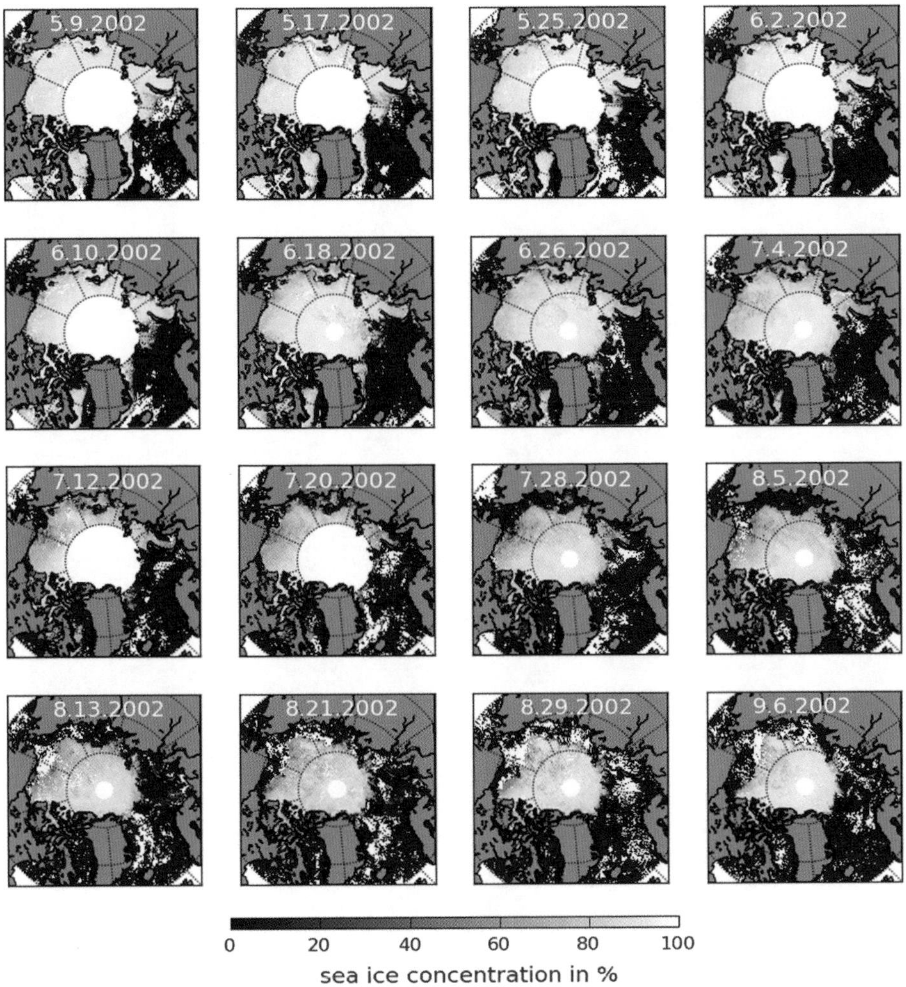
5.9.2002
5.17.2002
5.25.2002
6.2.2002
6.10.2002
6.18.2002
6.26.2002
7.4.2002
7.12.2002
7.20.2002
7.28.2002
8.5.2002
8.13.2002
8.21.2002
8.29.2002
9.6.2002
0
20
40
60
80
100
sea ice concentration in %

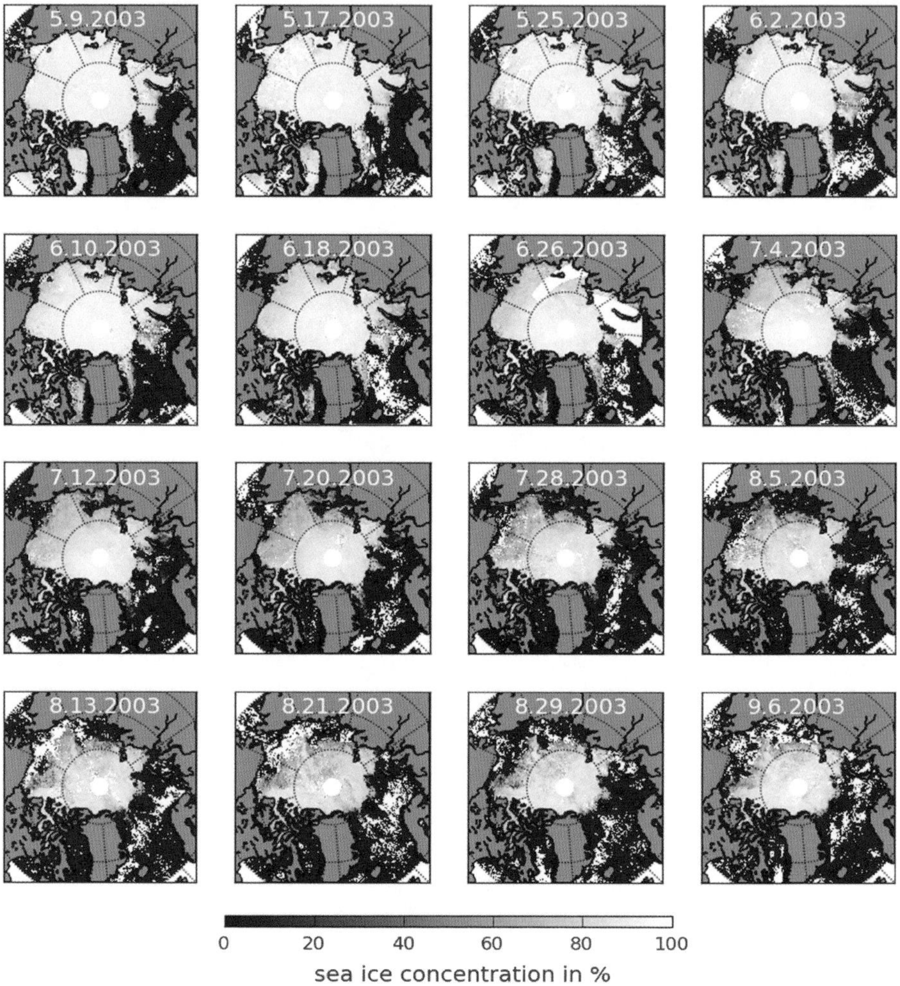
5.9.2003
5.17.2003
5.25.2003
6.2.2003
6.10.2003
6.18.2003
6.26.2003
7.4.2003
7.12.2003
7.20.2003
7.28.2003
8.5.2003
8.13.2003
8.21.2003
8.29.2003
9.6.2003
0
20
40
60
80
100
sea ice concentration in %

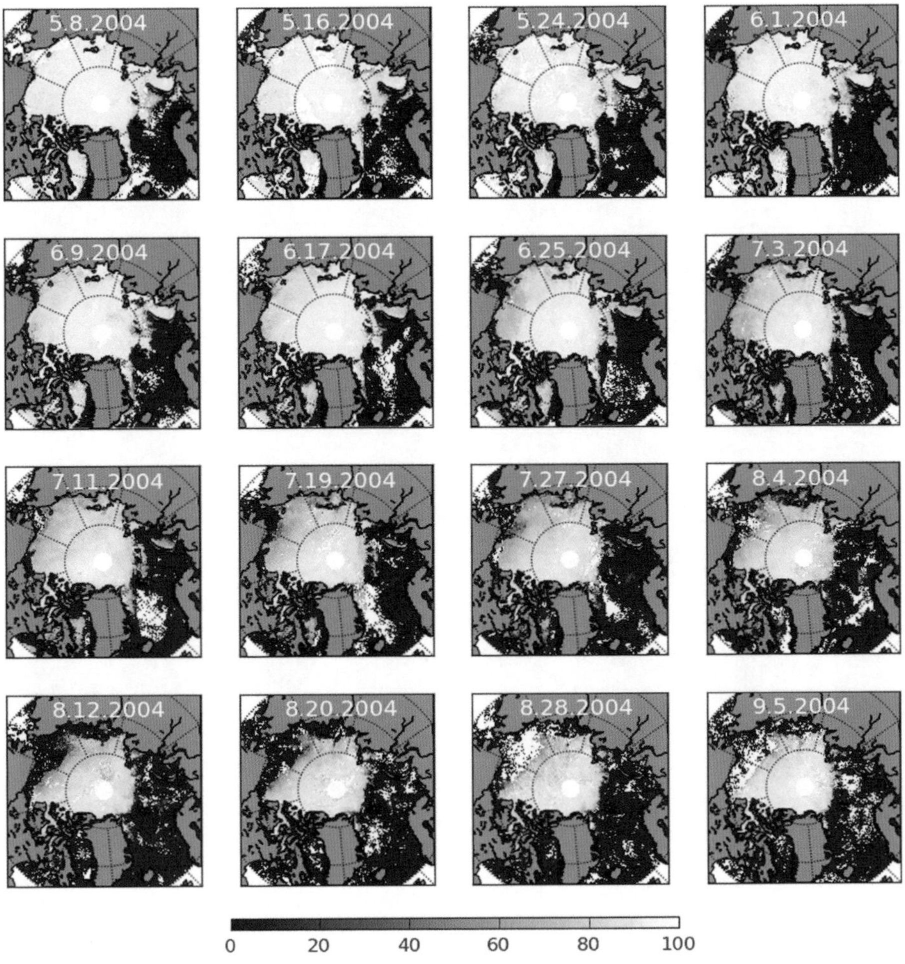
5.8.2004
5.16.2004
5.24.2004
6.1.2004
6.9.2004
6.17.2004
6.25.2004
7.3.2004
7.11.2004
7.19.2004
7.27.2004
8.4.2004
8.12.2004
8.20.2004
8.28.2004
9.5.2004
0
20
40
60
80
100
sea ice concentration in %

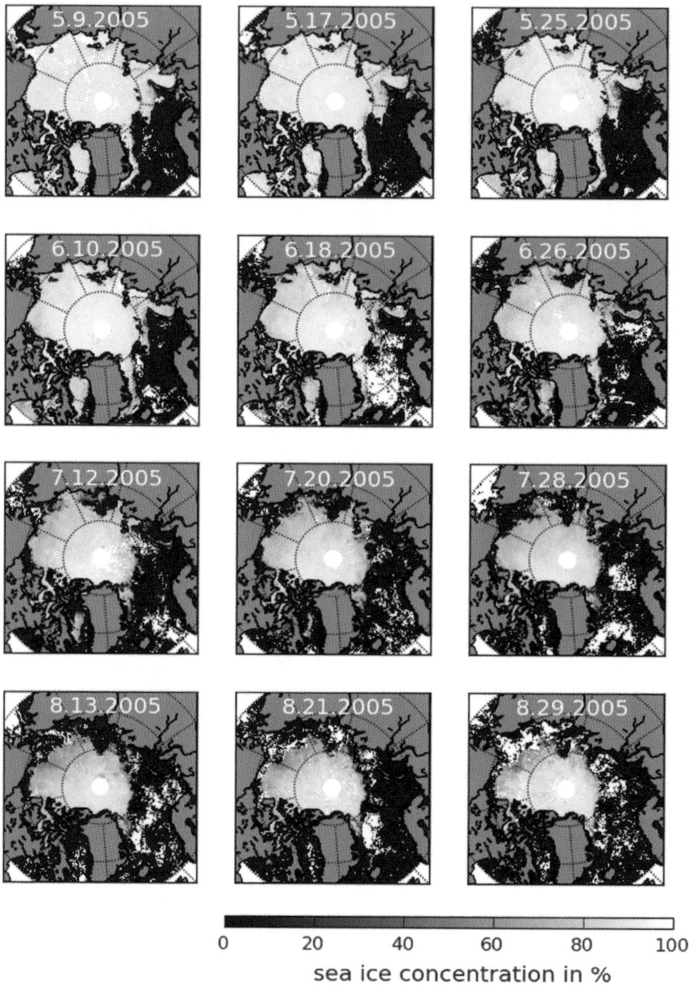
5.9.2005
5.17.2005
5.25.2005
6.2.2005
6.10.2005
6.18.2005
6.26.2005
7.4.2005
7.12.2005
7.20.2005
7.28.2005
8.5.2005
8.13.2005
8.21.2005
8.29.2005
9.6.2005
0
20
40
60
80
100
sea ice concentration in %

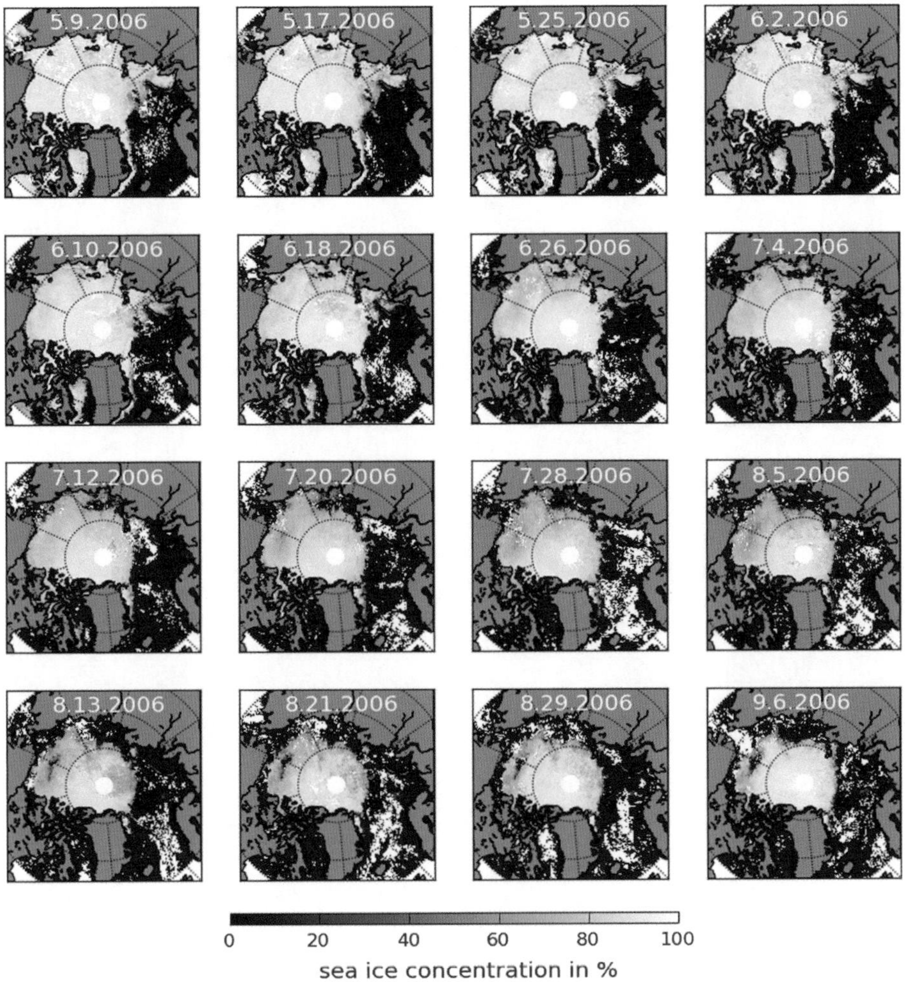
5.9.2006
5.17.2006
5.25.2006
6.2.2006
6.10.2006
6.18.2006
6.26.2006
7.4.2006
7.12.2006
7.20.2006
7.28.2006
8.5.2006
8.13.2006
8.21.2006
8.29.2006
9.6.2006
0
20
40
60
80
100
sea ice concentration in %

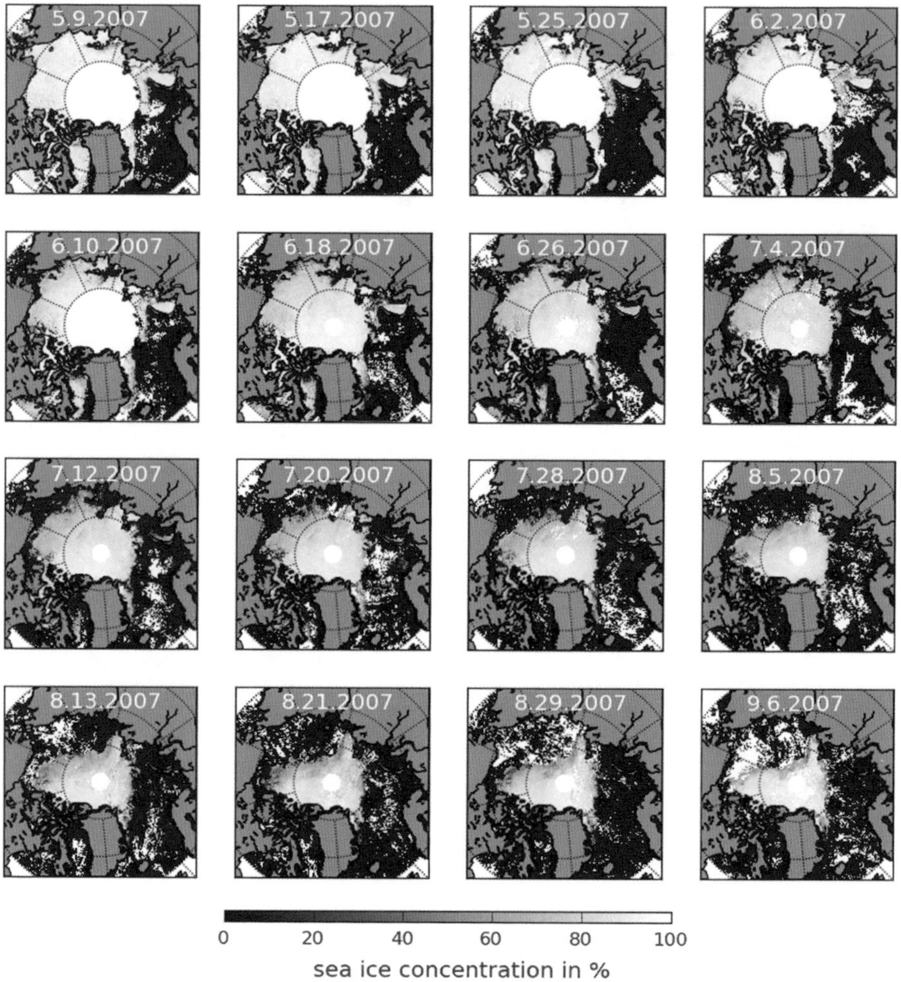
5.9.2007
5.17.2007
5.25.2007
6.2.2007
6.10.2007
6.18.2007
6.26.2007
7.4.2007
7.12.2007
7.20.2007
7.28.2007
8.5.2007
8.13.2007
8.21.2007
8.29.2007
9.6.2007
0
20
40
60
80
100
sea ice concentration in %

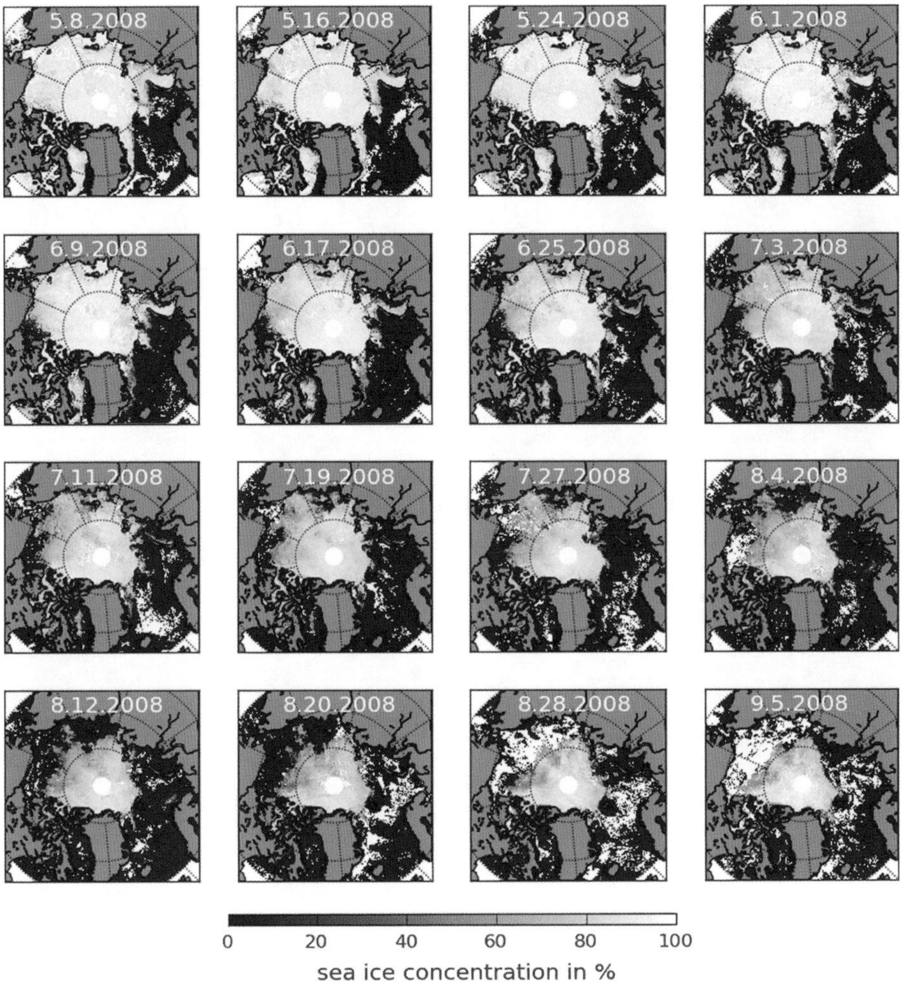
5.8.2008
5.16.2008
5.24.2008
6.1.2008
6.9.2008
6.17.2008
6.25.2008
7.3.2008
7.11.2008
7.19.2008
7.27.2008
8.4.2008
8.12.2008
8.20.2008
8.28.2008
9.5.2008
0
20
40
60
80
100
sea ice concentration in %

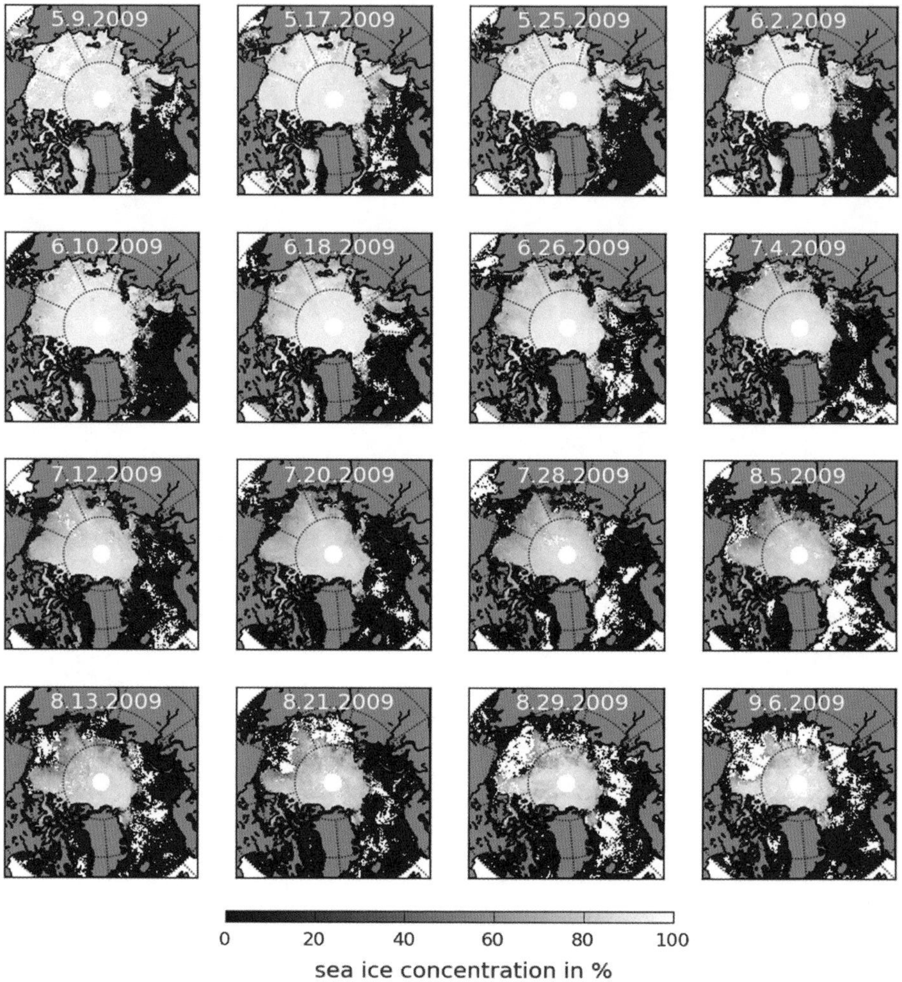
5.9.2009
5.17.2009
5.25.2009
6.2.2009
6.10.2009
6.18.2009
6.26.2009
7.4.2009
7.12.2009
7.20.2009
7.28.2009
8.5.2009
8.13.2009
8.21.2009
8.29.2009
9.6.2009
0
20
40
60
80
100
sea ice concentration in %

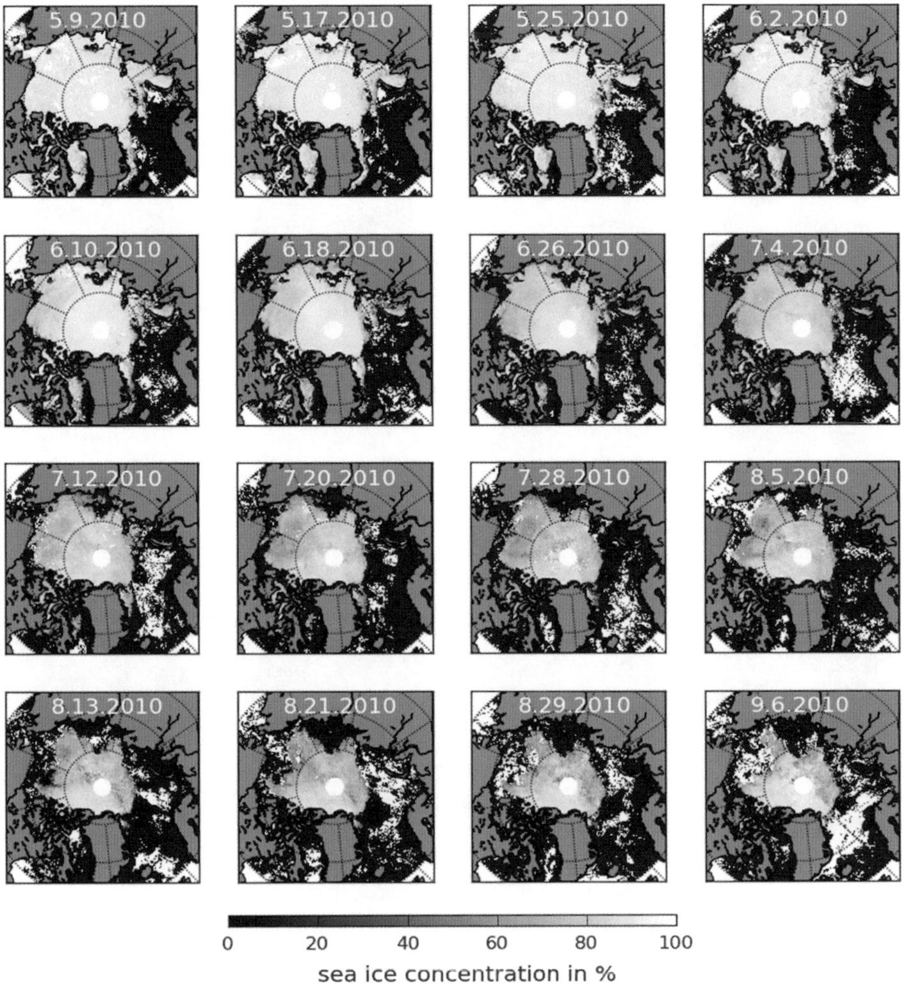
5.9.2010
5.17.2010
5.25.2010
6.2.2010
6.10.2010
6.18.2010
6.26.2010
7.4.2010
7.12.2010
7.20.2010
7.28.2010
8.5.2010
8.13.2010
8.21.2010
8.29.2010
9.6.2010
0
20
40
60
80
100
sea ice concentration in %

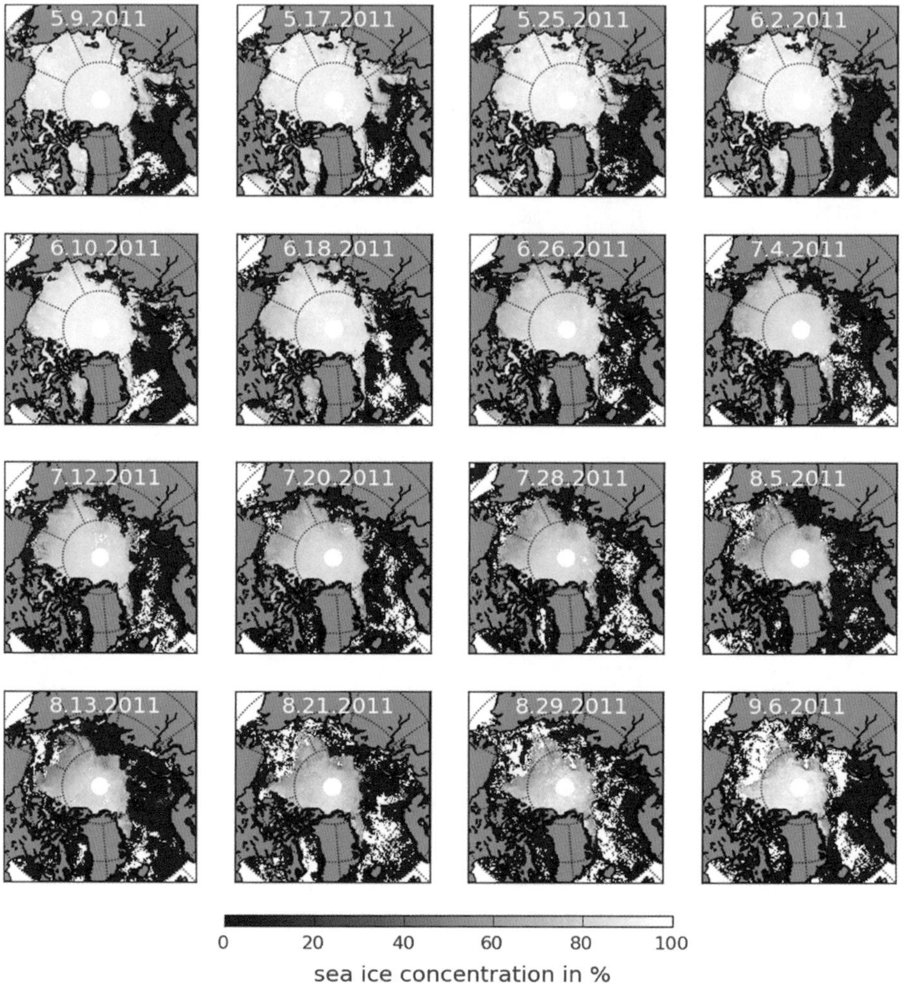

Fig. 6.1 Seasonal cycle of the sea ice concentration from MODIS satellite data for the Arctic for the years 2000–2011

A.3 MODIS Melt Ponds (Fig. 6.2)

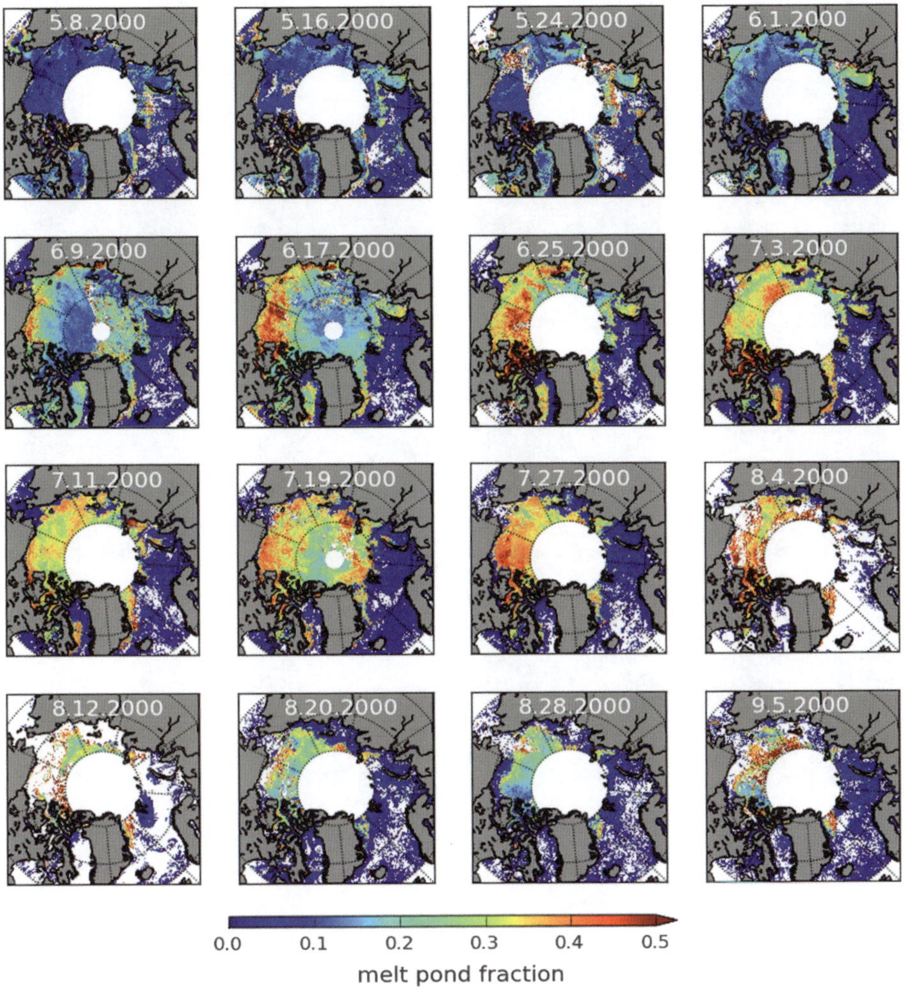

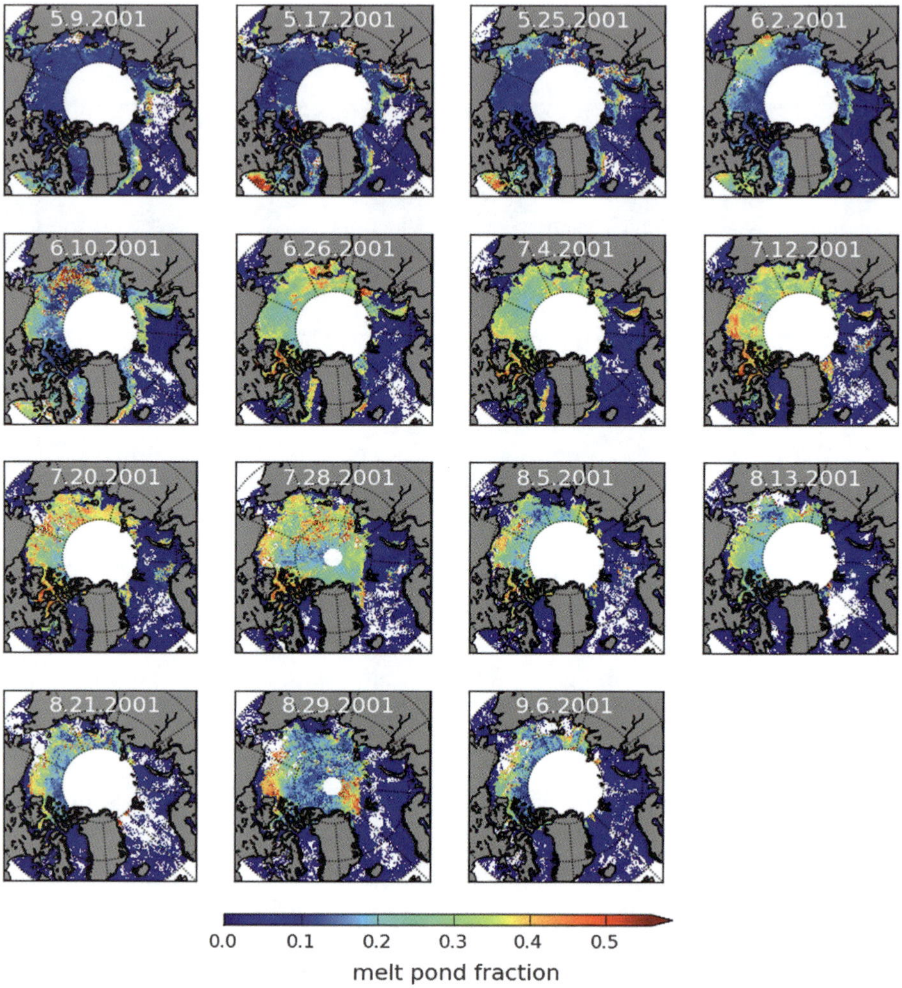
5.9.2001
5.17.2001
5.25.2001
6.2.2001
6.10.2001
6.26.2001
7.4.2001
7.12.2001
7.20.2001
7.28.2001
8.5.2001
8.13.2001
8.21.2001
8.29.2001
9.6.2001
0.0
0.1
0.2
0.3
0.4
0.5
melt pond fraction

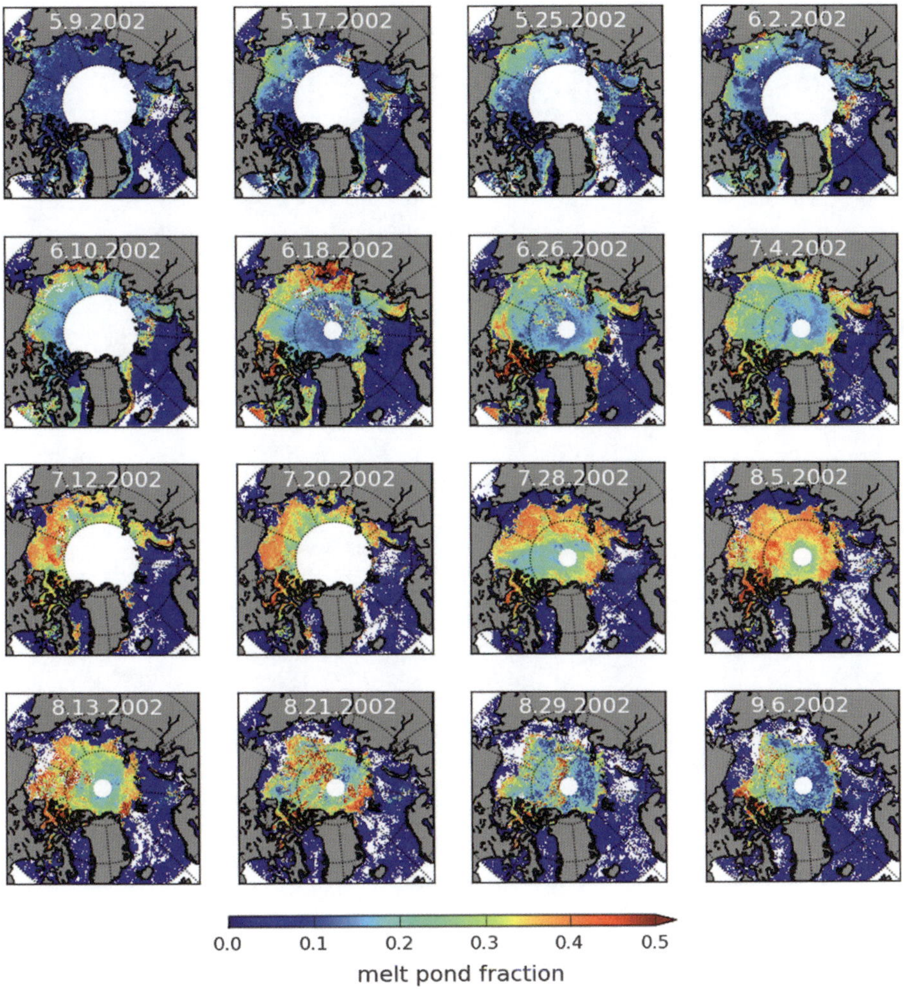
5.9.2002
5.17.2002
5.25.2002
6.2.2002
6.10.2002
6.18.2002
6.26.2002
7.4.2002
7.12.2002
7.20.2002
7.28.2002
8.5.2002
8.13.2002
8.21.2002
8.29.2002
9.6.2002
0.0
0.1
0.2
0.3
0.4
0.5
melt pond fraction

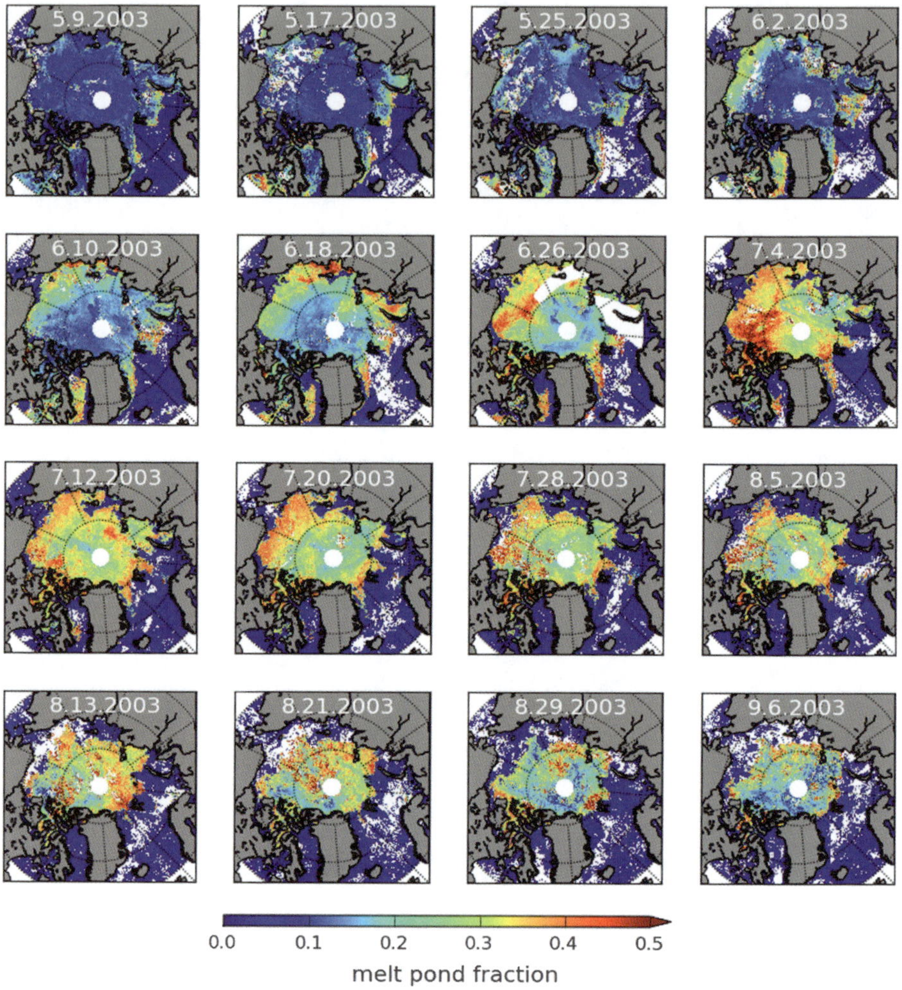
5.9.2003
5.17.2003
5.25.2003
6.2.2003
6.10.2003
6.18.2003
6.26.2003
7.4.2003
7.12.2003
7.20.2003
7.28.2003
8.5.2003
8.13.2003
8.21.2003
8.29.2003
9.6.2003
0.0
0.1
0.2
0.3
0.4
0.5
melt pond fraction

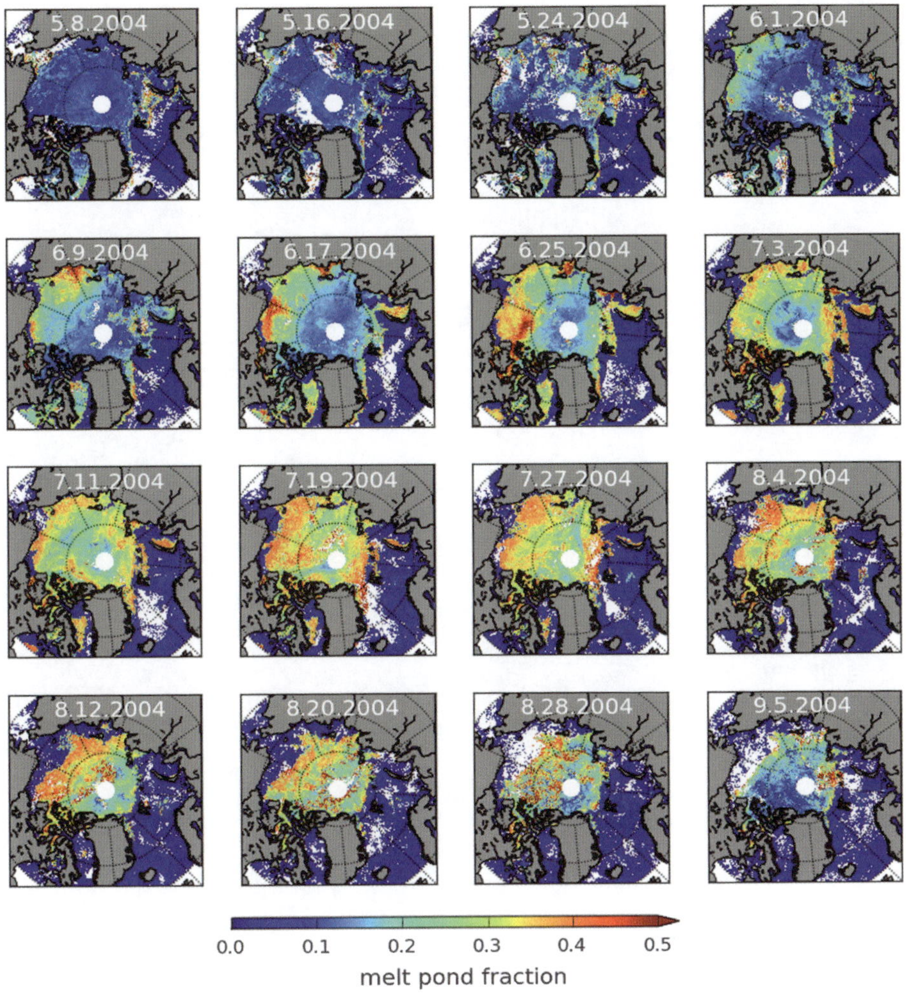
5.8.2004
5.16.2004
5.24.2004
6.1.2004
6.9.2004
6.17.2004
6.25.2004
7.3.2004
7.11.2004
7.19.2004
7.27.2004
8.4.2004
8.12.2004
8.20.2004
8.28.2004
9.5.2004
0.0
0.1
0.2
0.3
0.4
0.5
melt pond fraction

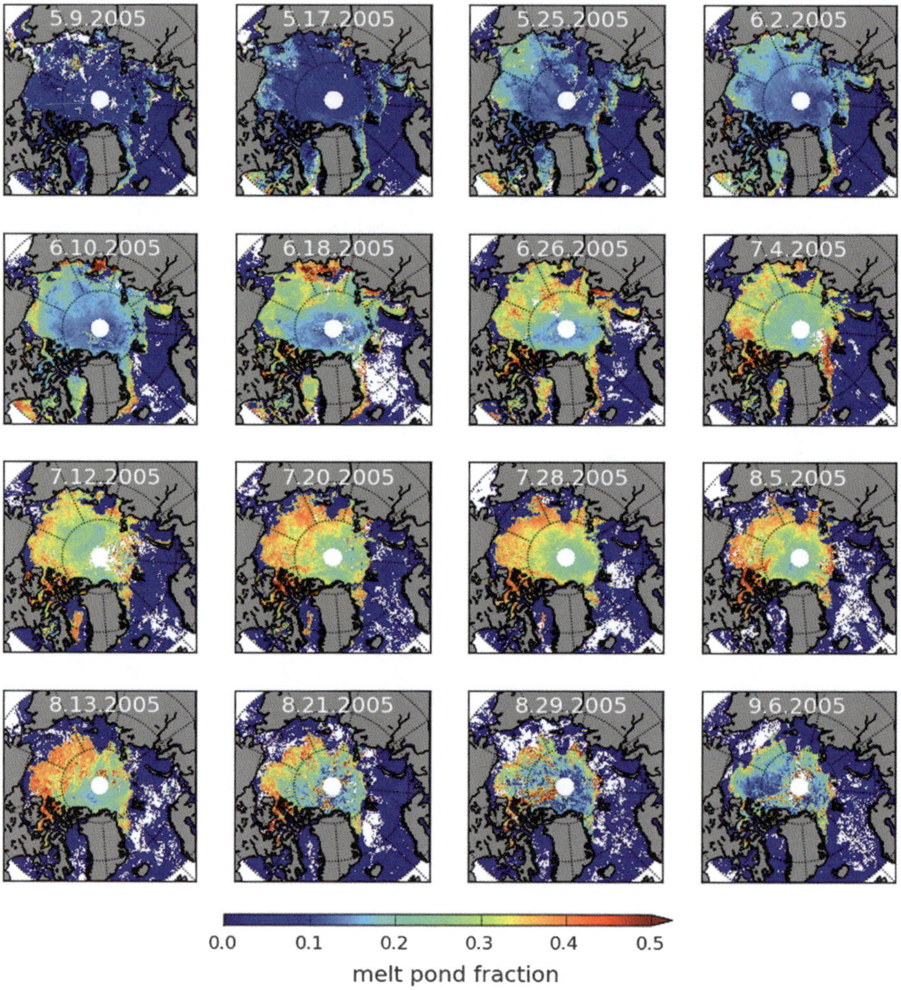
5.9.2005
5.17.2005
5.25.2005
6.2.2005
6.10.2005
6.18.2005
6.26.2005
7.4.2005
7.12.2005
7.20.2005
7.28.2005
8.5.2005
8.13.2005
8.21.2005
8.29.2005
9.6.2005
0.0
0.1
0.2
0.3
0.4
0.5
melt pond fraction

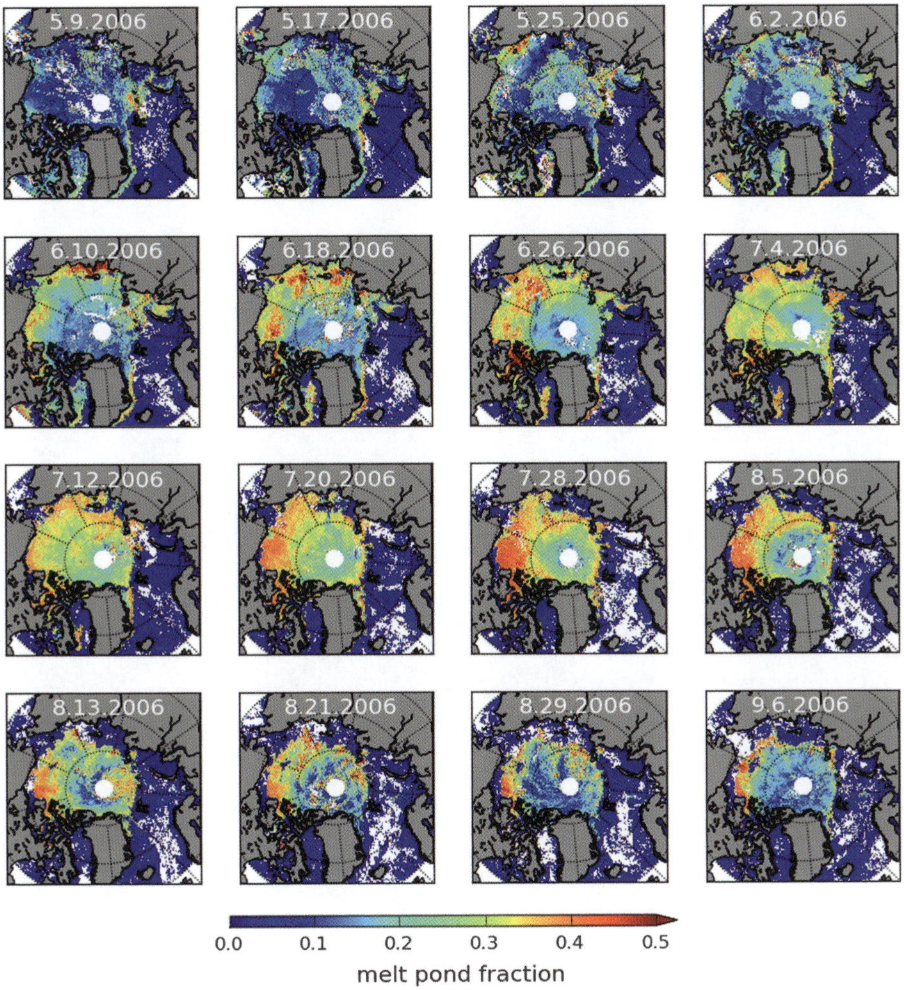
5.9.2006
5.17.2006
5.25.2006
6.2.2006
6.10.2006
6.18.2006
6.26.2006
7.4.2006
7.12.2006
7.20.2006
7.28.2006
8.5.2006
8.13.2006
8.21.2006
8.29.2006
9.6.2006
0.0
0.1
0.2
0.3
0.4
0.5
melt pond fraction

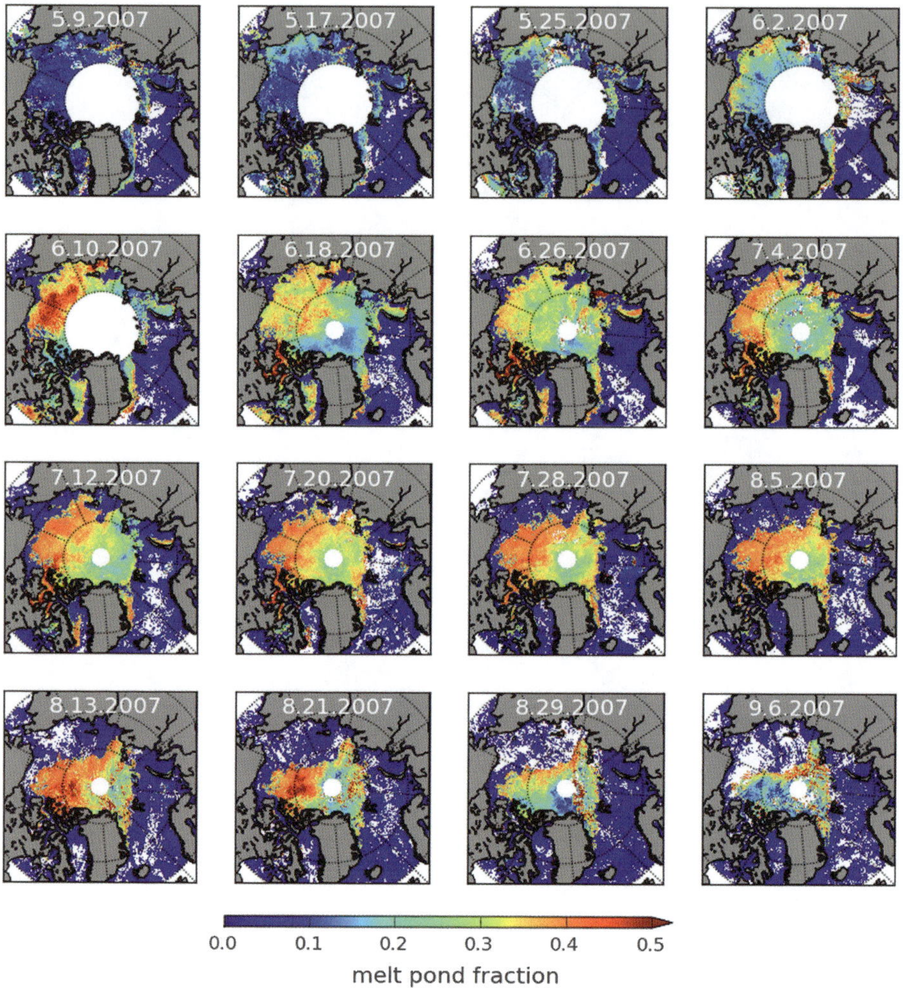
5.9.2007
5.17.2007
5.25.2007
6.2.2007
6.10.2007
6.18.2007
6.26.2007
7.4.2007
7.12.2007
7.20.2007
7.28.2007
8.5.2007
8.13.2007
8.21.2007
8.29.2007
9.6.2007
0.0
0.1
0.2
0.3
0.4
0.5
melt pond fraction

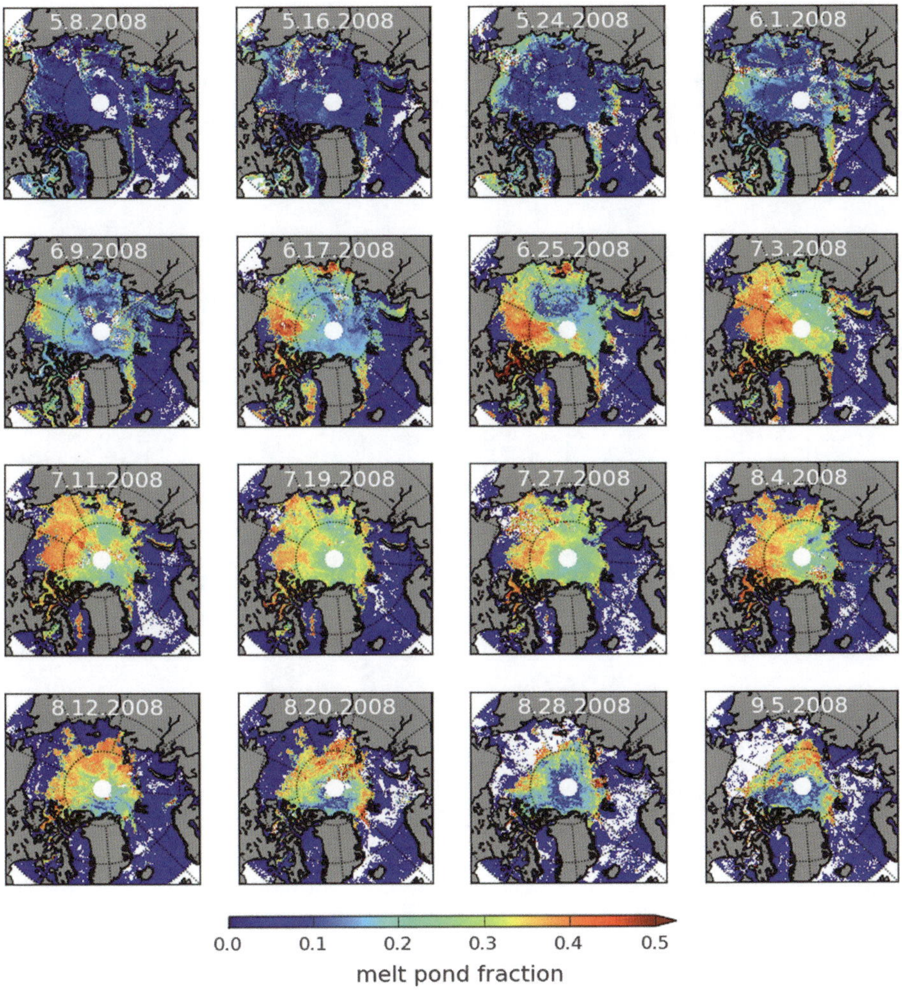
5.8.2008
5.16.2008
5.24.2008
6.1.2008
6.9.2008
6.17.2008
6.25.2008
7.3.2008
7.11.2008
7.19.2008
7.27.2008
8.4.2008
8.12.2008
8.20.2008
8.28.2008
9.5.2008
0.0
0.1
0.2
0.3
0.4
0.5
melt pond fraction

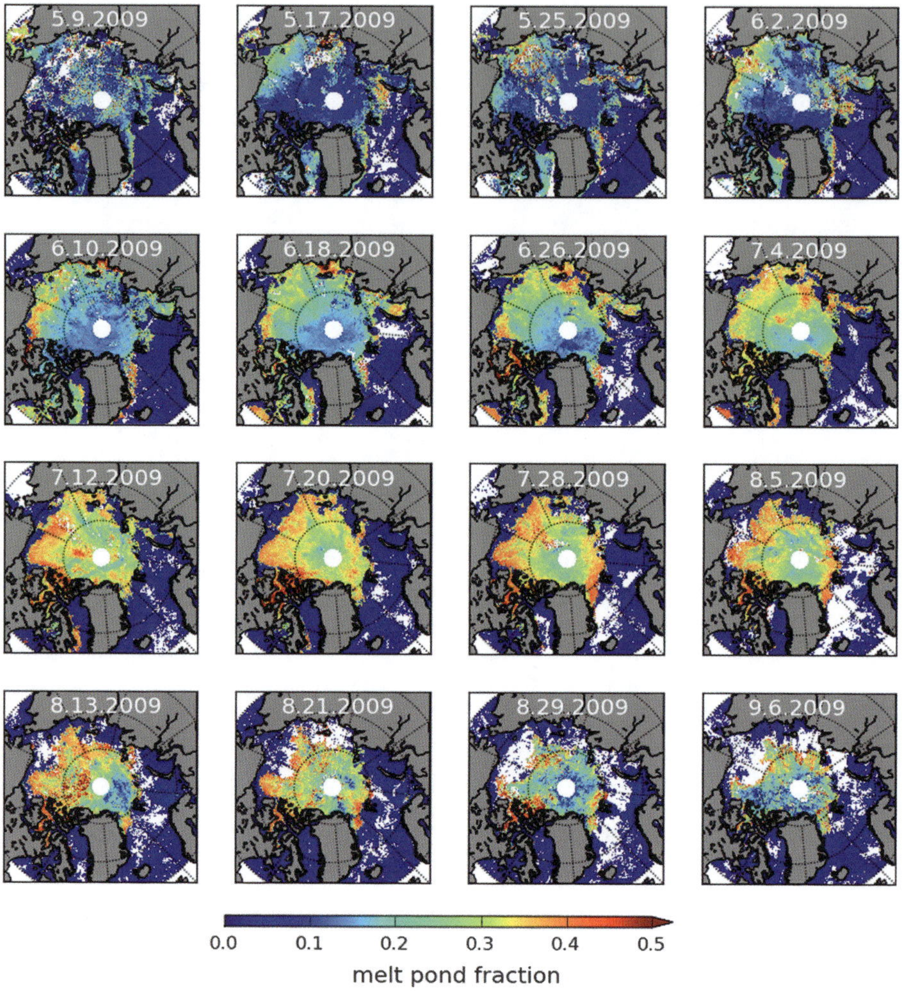
5.9.2009
5.17.2009
5.25.2009
6.2.2009
6.10.2009
6.18.2009
6.26.2009
7.4.2009
7.12.2009
7.20.2009
7.28.2009
8.5.2009
8.13.2009
8.21.2009
8.29.2009
9.6.2009
0.0
0.1
0.2
0.3
0.4
0.5
melt pond fraction

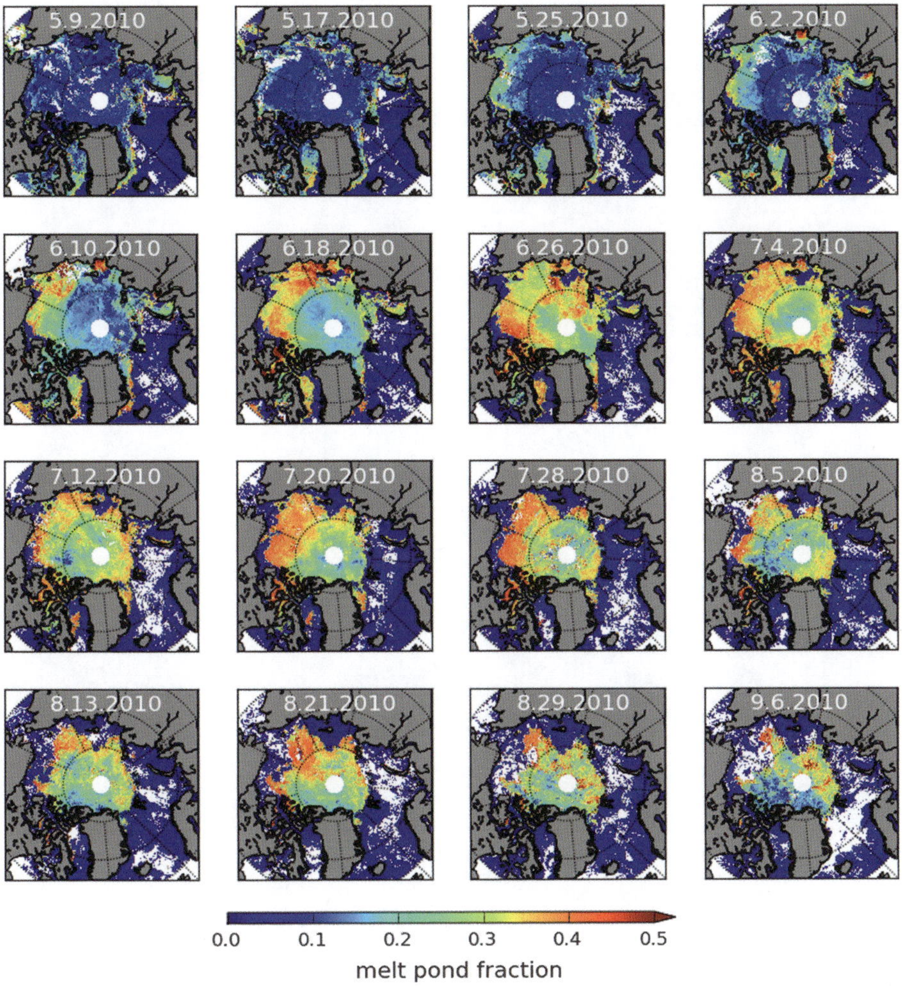
5.9.2010
5.17.2010
5.25.2010
6.2.2010
6.10.2010
6.18.2010
6.26.2010
7.4.2010
7.12.2010
7.20.2010
7.28.2010
8.5.2010
8.13.2010
8.21.2010
8.29.2010
9.6.2010
0.0
0.1
0.2
0.3
0.4
0.5
melt pond fraction

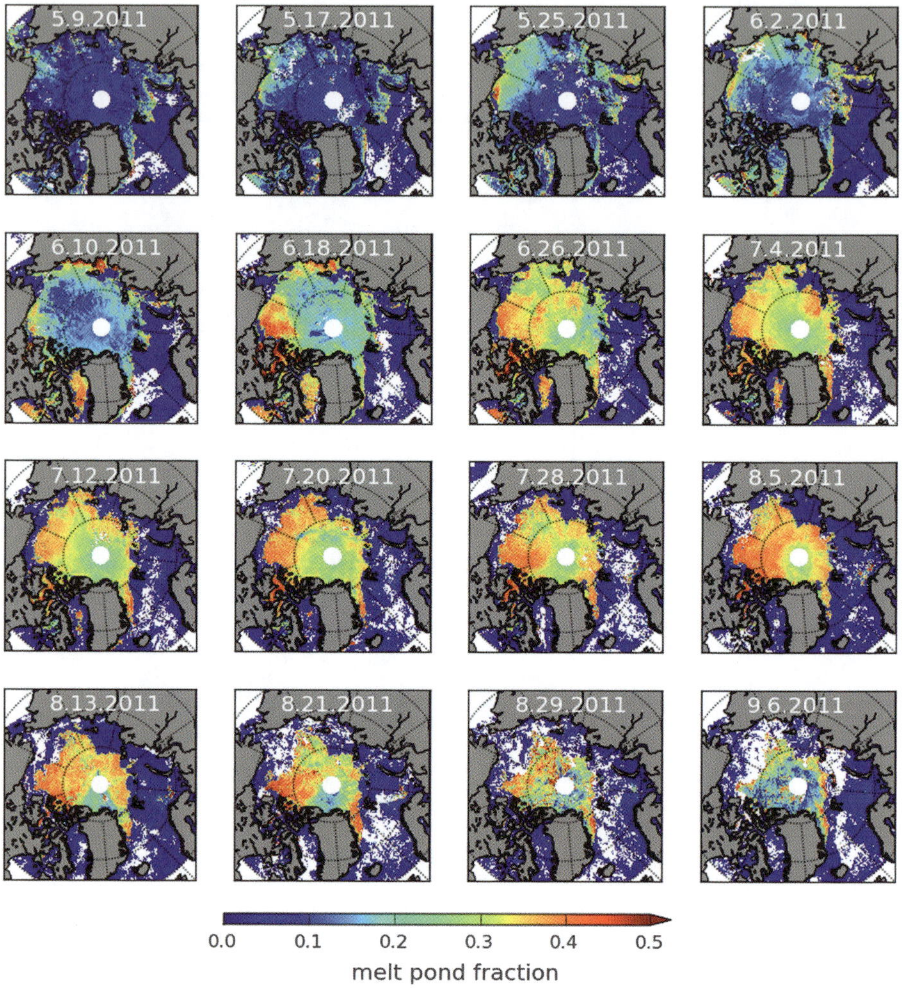

Fig. 6.2 Seasonal cycles of the melt pond fraction from MODIS satellite data for the Arctic for the years 2000–2011

Glossary

Technical terms regarding sea ice are partly literally cited from the WMO Sea Ice Nomenclature (1989). This document gives recommendations for future ice chart preparation using the international system of sea ice symbols.

absolute melt pond area → Relative melt pond fraction scaled with sea ice area.

albedo → Total albedo.

areal-averaged albedo Surface-based → total albedo values are weighted with the fraction of their corresponding surface component.

Arctic Ocean The Arctic Ocean is located in the northern hemisphere in the Arctic north polar region. The Arctic Ocean is almost completely surrounded by Eurasia and North America. In this study we use satellite date from 60°N northwards to cover the Arctic Ocean.

Aqua A NASA Earth Science satellite mission, part of the NASA-centered international Earth Observing System (EOS). Aqua is member of a group of satellites named the Afternoon Constellation, or sometimes the A-Train. Aqua was launched on May 4, 2002, and has six Earth-observing instruments on board, including the → MODIS sensor, and is collecting a variety of global data sets.

BFGS Broyden–Fletcher–Goldfarb–Shanno method; a method for solving nonlinear mathematical optimization problems.

BRDF Bidirectional reflectance distribution function. The BRDF is a four-dimensional function that defines how light is reflected at an opaque surface.

broadband albedo → Total albedo.

CloudSat CloudSat is a NASA Earth observation satellite, launched on April 28, 2006. It uses radar to measure the altitude and properties of clouds. CloudSat is like → Aqua and → CALIPSO part of the "A Train"-formation.

CALIPSO Cloud–Aerosol Lidar and Infrared Pathfinder Satellite Observations is a joint NASA and CNES environmental satellite, launched on April 28, 2006. Passive and active remote sensing Instruments monitor aerosols and clouds. CALIPSO is part of the "A Train", flying in formation with several other satellites (→ Aqua, Aura, → CloudSat and PARASOL).

A. Rösel, *Detection of Melt Ponds on Arctic Sea Ice with Optical Satellite Data*,
Hamburg Studies on Maritime Affairs 25, DOI 10.1007/978-3-642-37033-5,

crack Any fracture of fast ice, consolidated ice or a single floe with a width ranging from a few centimeters to 50 m and a length from several tens or hundreds of meters to several hundreds of kilometers.

first-year ice Sea ice of not more than one winter's growth, developing from young ice; thickness 30 cm to 2 m, and sometimes slightly more. May be subdivided into thin first-year ice/white ice, medium first-year ice and thick first-year ice.

drift ice Drift ice is ice that floats on the surface of the water in cold regions, as opposed to fast ice, which is attached to a shore. Usually drift ice is carried along by winds and sea currents. See also → pack ice.

fast ice Consolidated solid ice attached to the shore, to an ice wall or to an ice front. It forms by freezing to the shore of the ice cover forming in the coastal zone or as a result of freezing of drifting ice of any age category to the shore or fast ice. Vertical movement may be observed during tidal oscillations. It can be preserved without fracturing for two or more years transforming from first-year ice to multi-year ice and even shelf ice. The fast ice width can vary from several hundreds of meters to several hundreds of kilometers.

floe Any relatively flat piece of sea ice 20 m or more across.

HDF Hierarchical Data Format—http://www.hdfgroup.org/. File formats and libraries designed to store and organize large amounts of numerical data.

HOTRAX Healy-Oden Trans-Arctic Expedition 2005; This expedition was aimed to cross the Arctic Ocean. The U.S. Coast Guard Icebreaker Healy left Dutch Harbor, Alaska, on August 5, 2005, conducted various geophysical experiments and reached the North Pole on September 12. After the Pole, it took more than a week to get through 200 miles of heavy ice pack, reaching Tromsø, Norway on September 30, 2005.

ice-albedo feedback Ice-albedo feedback is a positive feedback climate process where a change in the area of snow covered land, ice caps, glaciers or sea ice alters the albedo. This change in albedo acts to reinforce the initial alteration in ice area. Cooling tends to increase ice cover and hence the albedo, reducing the amount of solar energy absorbed and leading to more cooling. Conversely, warming tends to decrease ice cover and hence the albedo, increasing the amount of solar energy absorbed, leading to more warming.

ice drift Displacement of ice floes and other ice features resulting from the impact of wind and currents including tidal currents and of forces transferred through the ice cover from other regions. The drift direction and velocity of a specific ice feature or ice cover area depends at any specific moment on the magnitude of the external forces, on the feature's characteristics (size, concentration and upper and lower surface roughness), on its position relative to the coastline and on the seabed relief.

Landsat 7 ETM+ The Earth observing sensor—the Enhanced Thematic Mapper Plus (ETM+)—on the Landsat 7 satellite was launched on April 15, 1999.

lead A more than 50 m wide rectilinear or wedge-shaped crack from several kilometers to several hundreds of kilometers in length. At below freezing temperatures, new, nilas and young ice forms at the surface of leads.

level ice Sea ice which has not been affected by deformation.

melt pond An accumulation melt water on ice, mainly due to melting snow, but in the more advanced stages also to the melting of ice. The initial stage consists of patches of melted snow.

MELTEX Aircraft campaign MELTEX "Impact of melt ponds on energy and momentum fluxes between atmosphere and sea ice", conducted by the → AWI in May and June 2008 over the Beaufort Sea.

mixed pixel A pixel that contains spectral signals of more than one surface type.

MODIS The Moderate Resolution Imaging Spectroradiometer is a key instrument aboard the → Terra and → Aqua satellites.

multi-year ice Old ice up to 3 m or more thick that has survived at least two summers' melt. Hummocks are even smoother than in second-year ice and attain a look of mounds and hills. The surface of multi-year ice fields in places not subject to deformations is also hillocky due to non-uniform multiple melting. The ice is almost salt-free. Its color, where bare, is usually blue. As a result of melting, round puddles appear at its surface in summer and a well-developed drainage system is formed.

Nelder–Mead-Simplex Algorithm (fmin) The Nelder–Mead-Simplex method or downhill simplex method or amoeba method is a commonly used nonlinear optimization technique. However, the this technique is a heuristic search method that can converge to non-stationary points on problems that can be solved by alternative methods.

NetCDF Network Common Data Form is a data format that support the creation, access, and sharing of array-oriented scientific data.

nilas A thin elastic crust of ice, easily bending on waves and swell and under pressure, thrusting in a pattern of interlocking "fingers" (finger rafting). Has a matt surface and is up to 10 cm in thickness. May be subdivided into dark nilas and light nilas.

pack ice Any ice at the sea surface except for fast ice and stamukhas regardless of its age, form, origin and other characteristics that has a possibility of movement (drift) under the action of winds, currents and tides. As a result of the dynamic processes (drift, divergence, convergence), the total and partial concentrations of drifting ice constantly change.

pixel-weight-mask A mask of grid cells with a given size that contains more than 50 % of the involved pixels.

Principal Component Analysis (PCA) PCA is a mathematical procedure that uses an orthogonal transformation to convert a set of observations of possibly correlated variables into a set of values of uncorrelated variables called principal components.

relative melt pond fraction Melt pond area relative to the sea ice surface.

ridge A comparatively rectilinear conglomeration of ice fragments formed by pressure at the contact line between ice floes, usually along earlier existing cracks and leads or at the boundary between ice floes of different age.

sea ice concentration Sea ice concentration is defined as the area of sea ice relative to the total at a given point in the ocean.

SLC, SLC-off mode On May 31, 2003 the scan line corrector (SLC), which compensates for the forward motion of the satellite, failed. From this point in time Landsat 7 ETM+ is acquiring image data in the SLC-off mode.

spectral albedo or reflectivity Spectral albedo is the fraction of incident radiation reflected by a surface at a specific wavelength.

spectral irradiance Irradiance is the power of electromagnetic radiation per unit area (radiative flux) incident on a surface. The SI units are watts per square meter ($W\,m^{-2}$). When this is done for radiation incident on a surface, it is called spectral irradiance, and has SI units $W\,m^{-3}$, or commonly $W\,m^{-2}\,nm^{-1}$.

spectral radiance Spectral radiance is a radiometric measure that describe the amount of radiation such as light that passes through or is emitted from a particular area, and falls within a given solid angle in a specified direction. They are used to characterize both emission from diffuse sources and reflection from diffuse surfaces. The unit is watts per steradian per square meter ($W\,sr^{-1}\,mm^{-2}$).

SPOT "Satellite Pour l'Observation de la Terre" is a high-resolution, optical imaging Earth observation satellite system operating from space. It is run by Spot Image based in Toulouse, France. SPOT 5 was launched May 4, 2002 with 2.5, 5 and 10 m capability, SPOT 6 and SPOT 7 launches are scheduled for 2012 and 2013, respectively.

surface albedo In this context always total or broadband albedo.

Terra A NASA Earth Science satellite mission, part international Earth Observing System (EOS). Terra was launched on December 18, 1999, and has five Earth-observing instruments on board, including the]→ MODIS sensor, and is collecting a variety of global data sets.

TOA Top-of-the-atmosphere reflectance values—the reflectance as seen by the sensor. This includes also atmospheric influences like aerosols and water vapor.

total albedo Synonym: wavelength-integrated albedo; a measure of the total solar energy absorbed by the surface.

References

1. Ackerman, Frey, Strabala, Liu, Gumley, Baum, Menzel: Discriminating clear-sky from cloud with modis - algorithm theoretical basis document. Tech. rep., Cooperative Institute for Meteorological Satellite Studies, University of Wisconsin - Madison (2010)
2. Agarwal, S., Moon, W., Wettlaufer, J.S.: Decadal to seasonal variability of arctic sea ice albedo. Geophysical Research Letters **38**, L20,504 (2011). DOI 10.1029/2011GL049109
3. Atkinson, P.M., Tatnall, A.R.L.: Introduction - neural networks in remote sensing. International Journal of Remote Sensing **18**, 699–709 (1997). DOI 10.1080/014311697218700
4. Bindschadler, R., Vornberger, P., Fleming, A., Fox, A., Mullins, J., Binnie, D., Paulsen, S.J., Brian, G., Gorodetzky, D.: The landsat image mosaic of Antarctica. Remote Sensing of Environment **112**, 4214–4226 (2008)
5. Birnbaum, G.e., Dierking, W.e., Hartmann, J.e., Lüpkes, C.e., Ehrlich, A.e., Garbrecht, T.e., Sellmann, M.e.: The campaign meltex with research aircraft "polar 5" in the arctic in 2008. Berichte zur Polar- und Meeresforschung/Reports on Polar and Marine Research **593**, 3–85 (2009)
6. Brandt, R.E., Warren, S.G., Worby, A.P., Grenfell, T.C.: Surface albedo of the Antarctic sea ice zone. Journal of Climate **18**, 3606–3622 (2005)
7. Cavalieri, D.J., Burns, B.A., Onstott, R.G.: Investigation of the effects of summer melt on the calculation of sea ice concentration using active and passive microwave data. Journal of Geophysical Research **95, No. C4**, 5359–5369 (1990)
8. Cavalieri, D.J., Markus, T., Hall, D.K., Gasiewski, A.J., Klein, M., Ivanoff, A.: Assessment of eos aqua amsr-e arctic sea ice concentrations using landsat-7 and airborne microwave imagery. IEEE Transactions on Geoscience and Remote Sensing **44**(11, Part 1), 3057–3069 (2006). DOI 10.1109/TGRS.2006.878445
9. Chander, G., Markham, B.L., Helder, D.L.: Summary of current radiometric calibration coefficients for landsat mss, tm, etm+, and eo-1 ali sensors. Remote Sensing of Environment **113**(5), 893–903 (2009). DOI: 10.1016/j.rse.2009.01.007
10. Comiso, J.C.: Ssm/i sea ice concentrations using the bootstrap algorithm. NASA Reference Publication **1380**, 1–50 (1995)
11. Comiso, J.C.: Large decadal decline of the arctic multiyear ice cover. Journal of Climate **25**, 1176–1193 (2012)
12. Comiso, J.C., Kwok, R.: Surface and radiative characteristics of the summer arctic sea ice cover from multisensor satellite observation. Journal of Geophysical Research **101, No. C12**, 28,397–28,416 (1996)
13. Curry, J.A., Schramm, J.L., Ebert, E.E.: Sea ice-albedo climate feedback mechanism. Journal of Climate **8**, 240–247 (1995)

A. Rösel, *Detection of Melt Ponds on Arctic Sea Ice with Optical Satellite Data*,
Hamburg Studies on Maritime Affairs 25, DOI 10.1007/978-3-642-37033-5,

14. Duarte, C.M., Lenton, T.M., Wadhams, P., Wassmann, P.: Abrupt climate change in the arctic. Nature Climate Change **2**, 60–62 (2012)
15. Ehn, J.K., Mundy, C.J., Barber, D.G., Hop, H., Rossnagel, A., Stewart, J.: Impact of horizontal spreading on light propagation in melt pond covered seasonal sea ice in the Canadian arctic. Journal of Geophysical Research **116**, C00G02 (2011). DOI 10.1029/2010JC006980
16. Eicken, H., Grenfell, T.C., Perovich, D.K., Richter-Menge, J.A., Frey, K.: Hydraulic controls of summer arctic pack ice albedo. Journal of Geophysical Research **109**, C08,007 (2004). DOI 10.1029/2003JC001989
17. Eicken, H., Krouse, H.R., Kadko, D., Perovich, D.K.: Tracer studies of pathways and rates of meltwater transport through arctic summer sea ice. Journal of Geophysical Research **107**, C10,8046 (2002)
18. Eisenman, I., Wettlaufer, J.S.: Nonlinear threshold behavior during the loss of arctic sea ice-albedo. PNAS **106**, 28–32 (2009). DOI 10.1073/pnas.0806887106
19. El Naggar, S., Garrity, C., Ramseier, R.: Sea ice meltpond morphology and size distribution as determined from line scan camera local measurements in the arctic. IAPSO Proceedings XXI General Assembly **19**, Honolulu, Hawaii, 5–12 August (1995)
20. Fetterer, F., Knowles, K., Meier, W., Savoie, M.: Sea ice index (2002, updated 2009). URL http://nsidc.org/data/g02135.html
21. Fetterer, F., Untersteiner, N.: Observations of melt ponds on arctic sea ice. Journal of Geophysical Research **103**, 24, 821–24, 835 (1998)
22. Fetterer, F., Wilds, S., Sloan, J.: Arctic sea ice melt pond statistics and maps, 1999–2001. Digital Media (ftp) (2008). URL http://nsidc.org/data/g02159.html
23. Frey, K.E., Perovich, D.K., Light, B.: The spatial distribution of solar radiation under a melting arctic sea ice cover. Geophysical Research Letters **38**, L22,501 (2011). DOI 10.1029/2011GL049421
24. GDAL Development Team: GDAL - Geospatial Data Abstraction Library, Version 1.8.1. Open Source Geospatial Foundation (2011). URL http://www.gdal.org
25. Gonzalez Vilas, L., Evangelos, S., Torres Palenzuela, J.M.: Neural network estimation of chlorophyll a from meris full resolution data for the coastal waters of galician rias (nw Spain). Remote Sensing of Environment **115**, 524–535 (2011)
26. Grenfell, T., Maykut, G.: The optical properties of ice and snow in the arctic basin. Journal of Glaciology **18**, 445–463 (1977)
27. Grenfell, T.C., Perovich, D.K.: Spectral albedos of sea ice and incident solar irradiance in the southern Beaufort sea. Journal of Geophysical Research **89**, 3573–3580 (1984)
28. Grenfell, T.C., Perovich, D.K.: Seasonal and spatial evolution of albedo in a snow-ice-land-ocean environment. Journal of Geophysical Research **109**(C1), 15 pp. (2004)
29. Grenfell, T.C., Perovich, D.K.: Incident spectral irradiance in the arctic basin during the summer and fall. Journal of Geophysical Research **113**, D12,117 (2008). DOI 10.1029/2007JD009418
30. Holland, M.M., Bitz, C.M., Tremblay, B.: Future abrupt reductions in the summer arctic sea ice. Geophysical Research Letters **33**, L23,503 (2006). DOI 10.1029/2006GL028024
31. Howell, S.E.L., Tivy, A., Yackel, J., Scharien, R.: Application of a seawinds/quikscat sea ice melt algorithm for assessing melt dynamics in the canadian arctic archipelago. Journal of geophysical research: Biogeosciences **111**(C7) (2006). DOI 10.1029/2005JC003193. URL http://tinyurl.sfx.mpg.de/staj
32. Itoh, M., Inoue, J., Shimada, K., Zimmermann, S., Kikuchi, T., Hutchings, J., McLaughlin, F., Carmack, E.: Acceleration of sea-ice melting due to transmission of solar radiation through ponded ice area in the arctic ocean: results of in situ observation from icebreakers in 2006 and 2007. Annals of Glaciology **52(57)**, 249–260 (2011)
33. Jones, E., Oliphant, T., Peterson, P., et al.: SciPy: Open source scientific tools for Python. electronic (2001). URL http://www.scipy.org/
34. Kramer, H.J.: Observation of the Earth and its environment: survey of missions and sensors. Springer Verlag (2002)

35. Kurtz, N.T., Markus, T., Farrell, S.L., Worthen, D.L., Boisvert, L.N.: Observations of recent arctic sea ice volume loss and its impact on ocean-atmosphere energy exchange and ice production. Journal of Geophysical Research **116**, C04,015 (2011). DOI 10.1029/2010JC006235
36. Kwok, R.: New high-resolution images of summer arctic sea ice. EOS **7**, 53–54 (2011)
37. Kwok, R., Untersteiner, N.: The thinning of arctic sea ice. Physics today **41**, 36–41 (2011)
38. Landsat 7 Science Data Users Handbook: Landsat 7 Science Data Users Handbook (2009)
39. Lenton, T.M., Held, H., Kriegler, E., Hall, J.W., Lucht, W., Rahmstorf, S., Schellnhuber, H.J.: Tipping elements in the earth's climate system. PNAS **115(6)**, 1786–1793 (2008)
40. Levermann, A., Bamber, J.L., Drijfhout, S., Ganopolski, A., Haeberli, W., Harris, N.R.P., Huss, M., Krüger, K., Lenton, T.M., Lindsay, R.W., Notz, D., Wadhams, P., Weber, S.: Potential climatic transitions with profound impact on Europe. Climatic Change (2011). DOI 10.1007/s10584-011-0126-5
41. Liu, Y., Ackerman, S.A., Maddux, B.C., Key, J.R., Frey, R.A.: Errors in cloud detection over the arctic using a satellite imager and implications for observing feedback mechanisms. Journal of Climate **23**, 1894–1907 (2009). DOI 10.1175/2009JCLI3386.1 10.1175/2009JCLI3386.1 10.1175/2009JCLI3386.1 10.1175/2009JCLI3386.1 10.1175/2009JCLI3386.1
42. Markus, T., Cavalieri, D.J.: An enhancement of the nasa team sea ice algorithm. IEEE Transactions on Geoscience and Remote Sensing **38(3)**, 1387–1389 (2000)
43. Markus, T., Cavalieri, D.J., Ivanoff, A.: The potential of using landsat 7 etm+ for the classification of sea-ice surface conditions during summer. Annals of Glaciology **34**, 415–419 (2002)
44. Markus, T., Cavalieri, D.J., Tschudi, M.A., Ivanoff, A.: Comparison of aerial video and landsat 7 data over ponded sea ice. Remote Sensing of Environment **86**, 458–469 (2003). DOI 10.1016/S0034-4257(03)00124-X
45. Markus, T., Stroeve, J.C., Miller, J.: Recent changes in arctic sea ice melt onset, freezeup, and melt season length. Journal of Geophysical Research **114** (2009). DOI 10.1029/2009JC005436
46. Maslanik, J., Drobot, S., Fowler, C., McPhee, G., Emery, W., Barry, R.: On the arctic climate paradox and the continuing role of atmospheric circulation in affecting sea ice conditions. Geophysical Research Letters **34**, L03,711 (2007). DOI 10.1029/2006GL028269
47. Maslanik, J., Stroeve, J., Fowler, C., Emery, W.: Distribution and trends in arctic sea ice age through spring 2011. Geophysical Research Letters **38**, L13,502 (2011). DOI 10.1029/2011GL047735
48. MODIS Level 1B Product User's Guide: MODIS Level 1B Product User's Guide. NASA/Goddard Space Flight Center (2006)
49. Morassutti, M.P., LeDrew, E.F.: Albedo and depth of melt ponds on sea-ice. International Journal of Climatology **16**, 817–838 (1996). DOI 10.1002/(SICI)1097-0088(199607)16:7<817::AID-JOC44>3.0.CO;2-5
50. Nicolaus, M., Gerland, S., Hudson, S.R., Hanson, S., Haapala, J., Perovich, D.K.: Seasonality of spectral albedo and transmittance as observed in the arctic transpolar drift in 2007. Journal of Geophysical Research **115**, C11,011 (2010). DOI 10.1029/2009JC006074
51. Nocedal, J., Wright, S.J.: Numerical Optimization (2nd ed.). Springer (2006)
52. Notz, D.: The future of ice sheets and sea ice: Between reversible retreat and unstoppable loss. PNAS **106**, 20,590–20,595 (2009). DOI 10.1073/pnas.0902356106
53. Overland, J., Bhatt, U., Key, J., Liu, Y., Walsh, J., Wang, M.: Temperature and clouds (2011). URL http://www.arctic.noaa.gov/repordcard/temperature_clouds.html
54. Perovich, D.K.: The optical properties of sea ice. CRREL Monograph **96-1**, 25 pp. (1996)
55. Perovich, D.K., Grenfell, T.C., Light, B., Elder, B.C., Harbeck, J., Polashenski, C., Tucker III, W.B., Stelmach, C.: Transpolar observations of the morphological properties of arctic sea ice-albedo. Journal of Geophysical Research. **114**, C00A04 (2009). DOI 10.1029/2008JC004892
56. Perovich, D.K., Grenfell, T.C., Light, B., Hobbs, P.V.: Seasonal evolution of the albedo of multiyear arctic sea ice. Journal of Geophysical Research. **107**(C10), 8044 (2002). DOI 10.1029/2000JC000438

57. Perovich, D.K., Jones, K.F., Light, B., Eicken, H., Markus, T., Stroeve, J., Lindsay, R.: Solar partitioning in a changing arctic sea-ice cover. Annals of Glaciology **52(57)**, 192–196 (2011)
58. Perovich, D.K., Light, B., Eicken, H., Jones, K.F., Runciman, K., Nghiem, S.V.: Increasing solar heating of the arctic ocean and adjacent seas, 1979–2005: Attribution and role in the ice-albedo feedback. Geophysical Research Letters **34**, L19,505 (2007). DOI 10.1029/2007GL031480
59. Perovich, D.K., Richter-Menge, J.A., Jones, K.F., Light, B., Elder, B.C., Polashenski, C., Laroche, D., Markus, T., Lindsay, R.: Arctic sea-ice melt in 2008 and the role of solar heating. Annals of Glaciology **52(57)**, 355–359 (2011)
60. Perovich, D.K., Tucker, W.B.I.: Arctic sea-ice conditions and distribution of solar radiation during summer. Annals of Glaciology **25**, 445–450 (1997)
61. Perovich, D.K., Tucker III, W.B., Ligett, K.A.: Aerial observations of the evolution of ice surface conditions during summer. Journal of Geophysical Research. **107**, 8048 (2002). DOI 10.1029/2000JC000449
62. Petty, W.G.: A First Course in Atmospheric Radiation, 2nd edn. Sundog Publishing (2006)
63. Rösel, A., Kaleschke, L.: Comparison of different retrieval techniques for melt ponds on arctic sea ice from landsat and modis satellite data. Annals of Glaciology **52(57)**, 185–191 (2011)
64. Rösel, A., Kaleschke, L.: Exceptional melt pond occurrence in the years 2007 and 2011 on the arctic sea ice archived from modis satellite data. Journal of Geophysical Research **117**, C05,018 (2012). DOI 10.1029/2011JC007869
65. Rösel, A., Kaleschke, L., Birnbaum, G.: Melt ponds on arctic sea ice determined from modis satellite data using an artificial neuronal network. The Cryosphere **6-2**, 431–446 (2012). DOI 10.5194/tcd-6-431-2012
66. Rumelhart, D.E., Hinton, G.E., Williams, R.J.: Learning representations by back-propagating errors. Nature **323(6088)**, 533–536 (1986)
67. Sankelo, P., Haapala, J., Heiler, I., Eero, R.: Melt pond formation and temporal evolution at the station tara during summer 2007. Polar Research **29**, 311–321 (2010). DOI 10.1111/j.1751-8369.2010.00161.x
68. Schweiger, A.J., Zhang, J., Lindsay, R.W., Steele, M.: Did unusually sunny skies help drive the record sea ice minimum of 2007? Geophysical Research Letters **35**, L10,503 (2008). DOI 10.1029/2008GL033463
69. Serreze, M.C.: Rethinking the sea-ice tipping point. Nature **471**, 47–48 (2011). DOI 10.1038/471047a
70. Serreze, M.C., Barrett, A.P., Cassano, J.J.: Circulation and surface controls on the lower tropospheric air temperature field of the arctic. Journal of Geophysical Research **116**, D07,104 (2011). DOI 10.1029/2010JD015127
71. Serreze, M.C., Maslanik, J.A., Scharfen, G.R., Barry, R.G., Robinson, D.A.: Interannual variations in snow melt over arctic sea ice and relationships to atmospheric forcings. Annals of Glaciology **17**, 327–331 (1993)
72. Shokr, M.E., Sinha, N.K.: Arctic sea ice microstructure observations relevant to microwave scattering. Arctic **47, No. 3**, 265–279 (1994)
73. Spreen, G., Kaleschke, L., Heygster, G.: Sea ice remote sensing using amsr-e 89-ghz channels. Journal of Geophysical Research **113**, C02S03 (2008). DOI 10.1029/2005JC003384
74. Steffen, K., Schweiger, A.: Nasa team algorithm for sea ice concentration retrieval from defense meteorological satellite program special sensor microwave imager: comparison with landsat satellite imagery. Journal of Geophysical Research **96**(C12), 21,971–87 (1991)
75. Stroeve, J., Serreze, M., Drobot, S., Gearheard, S., Holland, M., Maslanik, J., Meier, W., Scambos, T.: Arctic sea ice extent plummets in 2007. EOS, Transction, AGU **89**, 13–20 (2008)
76. Stroeve, J.C., Serreze, M.C., Kay, J.E., Holland, M.M., Meier, W.M., Barrett, A.P.: The arctic's rapidly shrinking sea ice cover: A research synthesis. Climatic Change (2011). DOI 10.1007/s10584-011-0101-1
77. Tietsche, S., Notz, D., Jungclaus, J.H., Marotzke, J.: Recovery mechanisms of arctic summer sea ice. Geophysical Research Letters **38**, L02,707 (2011). DOI 10.1029/2010GL045698

78. Tschudi, M., Curry, J., Maslanik, J.: Determination of areal surface-feature coverage in the beaufort sea using aircraft video data. Annals of Glaciology **25**, 434–438 (1997)
79. Tschudi, M., Curry, J., Maslanik, J.: Airborne observations of summertime surface features and their effect on surface albedo during fire/sheba. Journal of Geophysical Research. **106**(D14), 15, 335–15, 344 (2001)
80. Tschudi, M.A., Maslanik, J.A., Perovich, D.K.: Melt pond coverage on arctic sea ice from modis. In: Proceeding, Amer. Met. Soc. 8th Conf. on Polar Meteorology and Ocean, San Diego, CA, 8–14 January, 2005 (2005)
81. Tschudi, M.A., Maslanik, J.A., Perovich, D.K.: Derivation of melt pond coverage on arctic sea ice using modis observation. Remote Sensing of Environment **112**, 2605–2614 (2008). DOI 10.1016/j.rse.2007.12.009
82. Vermonte, E.F., Kotchenova, S.Y., Ray, J.P.: MODIS Surface Reflectance User's Guide. MODIS Land Surface Reflectance Science Computing Facility, version 1.2 edn. (2008). URL http://modis-sr.ltdri.org/products/MOD09_UserGuide_v1_3.pdf
83. Warren, S.G.: Optical properties of snow. Reviews of Geophysics and Space Physics **20**, 67–89 (1982)
84. WMO: WMO Sea-Ice Nomenclature. WMO/Omm/BMO, tp. 145, supplement no. 5 edn. (1989)
85. Wojciechowski, M.: Ffnet: Feed-forward neural network for python, (2011). URL http://ffnet.sourceforge.net/, access date: 20 October 2011
86. Xiong, X., Stamnes, K., Lubin, D.: Surface albedo over the arctic ocean derived from avhrr and its validation with sheba data. Journal of applied Meteorology **41**, 413–425 (2002)
87. Yackel, J.J., Barber, D.G.: Melt ponds on sea ice in the canadian archipelago 2. on the use of radarsat-1 synthetic aperture radar for geophysical inversion. Journal of Geophysical Research **105**(C9), 22,061–22,070 (2000). DOI 10.1029/2000JC900076. URL http://tinyurl.sfx.mpg.de/stag
88. Yackel, J.J., Barber, D.G., Hanesiak, J.M.: Melt ponds on sea ice in the canadian archipelago: 1. variability in morphological and radiative properties. Journal of Geophysical Research **105(C9)**, 22,049–22,060 (2000). DOI 10.1029/2000JC900076

About the International Max Planck Research School for Maritime Affairs at the University of Hamburg

The International Max Planck Research School for Maritime Affairs at the University of Hamburg was established by the Max Planck Society for the Advancement of Science, in co-operation with the Max Planck Institute for Foreign Private Law and Private International Law (Hamburg), the Max Planck Institute for Comparative Foreign Public Law and International Law (Heidelberg), the Max Planck Institute for Meteorology (Hamburg) and the University of Hamburg. The School's research is focused on the legal, economic, and geophysical aspects of the use, protection, and organization of the oceans. Its researchers work in the fields of law, economics, and natural sciences. The School provides extensive research capacities as well as its own teaching curriculum. Currently, the School has 22 Directors who determine the general work of the School, act as supervisors for dissertations, elect applicants for the School's PhD-grants, and are the editors of this book series:

Prof. Dr. Dr. h.c. mult. Jürgen Basedow is Director of the Max Planck Institute for Foreign Private Law and Private International Law; *President and Professor Monika Breuch-Moritz* is the President of the German Federal Maritime and Hydrographic Agency; *Prof. Dr. Dr. h.c. Peter Ehlers* is the Director ret. of the German Federal Maritime and Hydrographic Agency; *Prof. Dr. Dr. h.c. Hartmut Graßl* is Director emeritus of the Max Planck Institute for Meteorology; *Dr. Tatiana Ilyina* is the Leader of the Research Group "Ocean Biogeochemistry" at the Max Planck Institute for Meteorology in Hamburg; *Prof. Dr. Florian Jeßberger* is Head of the International and Comparative Criminal Law Division at the University of Hamburg; *Prof. Dr. Lars Kaleschke* is Junior Professor at the Institute of Oceanography of the University of Hamburg; *Prof. Dr. Hans-Joachim Koch* is Director emeritus of the Seminar of Environmental Law at the University of Hamburg; *Prof. Dr. Robert Koch* is Director of the Institute of Insurance Law at the University of Hamburg; *Prof. Dr. Doris König* is the President of the Bucerius Law School; *Prof. Dr. Rainer Lagoni* is Director emeritus of the Institute of Maritime Law and the Law of the Sea at the University of Hamburg; *Prof. Dr. Gerhard Lammel* is Senior Scientist and Lecturer at the Max Planck Institute for Chemistry,

A. Rösel, *Detection of Melt Ponds on Arctic Sea Ice with Optical Satellite Data*, Hamburg Studies on Maritime Affairs 25, DOI 10.1007/978-3-642-37033-5,

Mainz; *Prof. Dr. Ulrich Magnus* is Managing Director of the Seminar of Foreign Law and Private International Law at the University of Hamburg; *Prof. Dr. Peter Mankowski* is Director of the Seminar of Foreign and Private International Law at the University of Hamburg; *Prof. Stefan Oeter* is Managing Director of the Institute for International Affairs at the University of Hamburg; *Prof. Dr. Marian Paschke* is Managing Director of the Institute of Maritime Law and the Law of the Sea at the University of Hamburg; *PD Dr. Thomas Pohlmann* is Senior Scientist at the Centre for Marine and Climate Research and Member of the Institute of Oceanography at the University of Hamburg; *Dr. Uwe A. Schneider* is Assistant Professor at the Research Unit Sustainability and Global Change of the University of Hamburg; *Prof. Dr. Detlef Stammer* is Professor in Physical Oceanography and Remote Sensing at the Institute of Oceanography of the University of Hamburg; *Prof. Dr. Jürgen Sündermann* is Director emeritus of the Centre for Marine and Climate Research at the University of Hamburg; *Prof. Dr. Rüdiger Wolfrum* is Director emeritus at the Max Planck Institute for Comparative Foreign Public Law and International Law and a judge at the International Tribunal for the Law of the Sea; *Prof. Dr. Wilfried Zahel* is Professor emeritus at the Centre for Marine and Climate Research of the University of Hamburg.

At present, *Prof. Dr. Dr. h.c. Jürgen Basedow* and *Prof. Dr. Ulrich Magnus* serve as speakers of the Research School.

Printed by Printforce, the Netherlands